Great Scientific Ideas That Changed the World
Parts I–III
Steven L. Goldman, Ph.D.

PUBLISHED BY:

THE TEACHING COMPANY
4151 Lafayette Center Drive, Suite 100
Chantilly, Virginia 20151-1232
1-800-TEACH-12
Fax—703-378-3819
www.teach12.com

Copyright © The Teaching Company, 2007

Printed in the United States of America

This book is in copyright. All rights reserved.

Without limiting the rights under copyright reserved above,
no part of this publication may be reproduced, stored in
or introduced into a retrieval system, or transmitted,
in any form, or by any means
(electronic, mechanical, photocopying, recording, or otherwise),
without the prior written permission of
The Teaching Company.

Steven L. Goldman, Ph.D.

Departments of Philosophy and History, Lehigh University

Steven L. Goldman has degrees in physics (B.Sc., Polytechnic University of New York) and philosophy (M.A., Ph.D., Boston University) and, since 1977, has been the Andrew W. Mellon Distinguished Professor in the Humanities at Lehigh University. He has a joint appointment in the departments of philosophy and history because his teaching and research focus on the history, philosophy, and social relations of modern science and technology. Professor Goldman came to Lehigh from the philosophy department at the State College campus of Pennsylvania State University, where he was a co-founder of one of the first U.S. academic programs in science, technology, and society (STS) studies. For 11 years (1977–1988), he served as director of Lehigh's STS program and was a co-founder of the National Association of Science, Technology and Society Studies. Professor Goldman has received the Lindback Distinguished Teaching Award from Lehigh University and a Book-of-the-Year Award for a book he co-authored (another book was a finalist and translated into 10 languages). He has been a national lecturer for Sigma Xi—the scientific research society—and a national program consultant for the National Endowment for the Humanities. He has served as a board member or as editor/advisory editor for a number of professional organizations and journals and was a co-founder of Lehigh University Press and, for many years, co-editor of its Research in Technology Studies series.

Since the early 1960s, Professor Goldman has studied the historical development of the conceptual framework of modern science in relation to its Western cultural context, tracing its emergence from medieval and Renaissance approaches to the study of nature through its transformation in the 20th century. He has published numerous scholarly articles on his social-historical approach to medieval and Renaissance nature philosophy and to modern science from the 17th to the 20th centuries and has lectured on these subjects at conferences and universities across the United States, in Europe, and in Asia. In the late 1970s, the professor began a similar social-historical study of technology and technological innovation since the Industrial Revolution. In the 1980s, he published a series of articles on innovation as a socially driven process and on the role played in that process by the knowledge created by scientists and engineers. These articles led to participation in science and technology policy initiatives of the federal government, which in turn led to extensive research and numerous article and book publications through the 1990s on emerging synergies that were transforming relationships among knowledge, innovation, and global commerce.

Professor Goldman is the author of two previous courses for The Teaching Company, *Science in the Twentieth Century: A Social Intellectual Survey* (2004) and *Science Wars: What Scientists Know and How They Know It* (2006).

Table of Contents
Great Scientific Ideas That Changed the World

Professor Biography i
Course Scope 1
Lecture One   Knowledge, Know-How, and Social Change 5
Lecture Two   Writing Makes Science Possible 10
Lecture Three   Inventing Reason and Knowledge 15
Lecture Four   The Birth of Natural Science 20
Lecture Five   Mathematics as the Order of Nature 25
Lecture Six   The Birth of Techno-Science 30
Lecture Seven   Universities Relaunch the Idea
                        of Knowledge 35
Lecture Eight   The Medieval Revolution in Know-How 40
Lecture Nine   Progress Enters into History 45
Lecture Ten   The Printed Book—Gutenberg to Galileo 51
Lecture Eleven   Renaissance Painting and Techno-Science 56
Lecture Twelve   Copernicus Moves the Earth 61
Lecture Thirteen   The Birth of Modern Science 67
Lecture Fourteen   Algebra, Calculus, and Probability 71
Lecture Fifteen   Conservation and Symmetry 76
Lecture Sixteen   Instruments as Extensions of the Mind 82
Lecture Seventeen   Time, Change, and Novelty 88
Lecture Eighteen   The Atomic Theory of Matter 93
Lecture Nineteen   The Cell Theory of Life 98
Lecture Twenty   The Germ Theory of Disease 103
Lecture Twenty-One The Gene Theory of Inheritance 108
Lecture Twenty-Two Energy Challenges Matter 113
Lecture Twenty-Three Fields—The Immaterial Becomes Real 119
Lecture Twenty-Four Relationships Become Physical 124
Lecture Twenty-Five Evolution as Process Science 129

Table of Contents
Great Scientific Ideas That Changed the World

Lecture Twenty-Six	Statistical Laws Challenge Determinism	135
Lecture Twenty-Seven	Techno-Science Comes of Age	141
Lecture Twenty-Eight	Institutions Empower Innovation	146
Lecture Twenty-Nine	The Quantum Revolution	152
Lecture Thirty	Relativity Redefines Space and Time	157
Lecture Thirty-One	Reconceiving the Universe, Again	162
Lecture Thirty-Two	The Idea behind the Computer	167
Lecture Thirty-Three	Three Faces of Information	172
Lecture Thirty-Four	Systems, Chaos, and Self-Organization	177
Lecture Thirty-Five	Life as Molecules in Action	183
Lecture Thirty-Six	Great Ideas, Past and Future	188
Timeline		194
Glossary		204
Biographical Notes		214
Bibliography		234

Great Scientific Ideas That Changed the World

Scope:

It is easy to fall into one of two traps in dealing with ideas: either to dismiss them as abstractions and, thus, of less consequence than concrete things, such as swords, plowshares, and factories, or to glorify them *as* abstractions, as creative inventions of the mind, and thus, praiseworthy independent of any practical consequences whatsoever. Ideas are, nevertheless, as concrete as swords and plowshares because they are always tied to a concrete context of values, actions, beliefs, artifacts, and institutions out of which they arise and on which they *may* act. The concreteness of ideas derives from their being produced not only *within* a particular cultural context but *out* of that context, and it is because ideas are produced out of a particular context that ideas are able to influence and even to reshape that context. Treating ideas out of context, then, treating them as if their existence were, in principle, independent of any particular context, deeply distorts the reality of ideas and obscures their power to affect the world.

Ideas and their contexts interact in complex, *mutually* influential ways such that the resultant effect on society of introducing a new idea is unpredictable. The evolution of the Internet from a modest computer networking project funded by the U.S. Department of Defense to a global technology transforming commerce, industry, politics, warfare, communication, education, entertainment, and research illustrates the unpredictability of the idea-social context interaction. The still-unfolding consequences of a small number of innovative ideas introduced to solve technical problems posed by enabling different kinds of computers in different locations to share information in real time continue to surprise, confound, and disturb us!

Unpredictable though it may be, however, for 200 years now, the interaction of science and technology with society has been the primary driver of social and cultural change, first in the West, then globally and at an accelerating rate. During this period, social and personal values and relationships; social, political, and economic institutions; and cultural values and activities have changed and continue to change almost beyond recognition by our great-grandparents. What is it that has enabled such deep transformations of ways of life that have been entrenched for centuries and even millennia?

Certainly, we can identify artifacts—the telephone, the automobile, airplanes, television, the computer—that *appear* to be causes of social change. But identifying artifacts does not reach down to the *causes* of innovation itself, nor does it expose those features of the sociocultural infrastructure that enable innovations to be causes of social change. Artifacts, in spite of their high visibility, are symptoms of causes at work; they are not themselves causes. It is not television or automobiles or the Internet that have changed society. Instead, forces at work within the network of relationships that we call society are causing television and automobiles and the Internet to take the changing forms that they take. One of these forces is ideas, explicitly in the case of new scientific ideas and implicitly in the case of ideas in the past that have been internalized selectively by society, thereby shaping both the sociocultural infrastructure and the lines along which it is vulnerable to change.

The objective of this course is to explore scientific ideas that have played a formative role in determining the infrastructure of modern life through a process of sociocultural selection. But we shall interpret the term *scientific idea* broadly. There is, after all, no sharp distinction between ideas that are classified as scientific and those that are classified as philosophical or mathematical or even between scientific ideas and political, religious, or aesthetic ideas. Alfred North Whitehead, for example, famously linked the emergence of modern science in the Christian West to Judaeo-Christian monotheism: to the belief in a single, law-observing creator of the Universe.

The idea that there are laws of nature at least *seems* to reflect a political idea, while there can be no doubt that mathematical and aesthetic ideas were central to the 17th-century Scientific Revolution. Furthermore, distinguishing science and technology is fuzzy, too, especially since the second half of the 19th century, when scientific knowledge and technological innovation were systematically coupled in industrial, academic, and government research laboratories.

With this in mind, we will begin our discussion of influential scientific ideas with the invention of writing, which may not seem a scientific idea at all. There is, nevertheless, a profound idea underlying the invention of writing, and a controversial one, as reflected in Socrates's argument *against* writing in Plato's dialogue *Phaedrus*. Writing is also a technology, of course, and thus, serves as an initial example of how technologies embody ideas that we tend to ignore because our attention is almost always drawn to *what* technologies do, to *how* they do it, and to what the consequences are of what they do.

By the time of the earliest written records that have been discovered so far, humans already had embodied, through their invention of a breathtaking range of physical, social, and cultural "technologies," an equally breathtaking range of ideas implicit in those technologies. Lecture One looks back at what humans had accomplished in the way of know-how by the 4th millennium B.C.E., while Lecture Two discusses the invention of writing and the spread of writing systems and texts from about 3500 B.C.E. to the beginning of classical antiquity, circa 500 B.C.E.

Between approximately 500 B.C.E. and 300 B.C.E., Greek philosophers developed highly specific concepts of knowledge, reason, truth, nature, mathematics, knowledge of nature, and the mathematical basis of knowledge of nature in ways that continue to inform the practice of science to the present day. Lectures Three through Five are devoted to these ideas and their legacy. Lecture Six discusses the first appearance in Western history, perhaps in world history, of the idea of techno-science, that is, of technology derived from theoretical knowledge rather than from practical know-how. This was largely a Greek idea that was applied in the context of the rising Roman Empire, and the lecture describes selected Roman-era technologies that had an influence on the rise of modern science and engineering.

Bridging the ancient and early modern eras, Lectures Seven through Eleven explore the idea of the university and its role as a progenitor of modern science; medieval machinery and Europe's first "industrial revolution"; and the Renaissance ideas of progress, of the printed book, and of mathematics as the language of nature. All these ideas are obviously seminal for science as we know it, but they are also, if less obviously, seminal for the rise of modern engineering and the form of modern technological innovation.

Lecture Twelve discusses Copernicus's idea of a moving Earth, the cultural consequences of that idea, and its subsequent evolution as a modern scientific astronomical theory. This serves as a lead-in to Lectures Thirteen through Seventeen, which explore foundational ideas of modern science, among them, the idea of method; new mathematical ideas, such as algebra and the calculus; ideas of conservation and symmetry; and the invention of new instruments that extended the mind rather than the senses and forced a new conception of knowledge.

Lectures Eighteen through Twenty-Eight explore 19th-century scientific ideas that remain profound social, cultural, and intellectual, as well as scientific, influences. These include the idea of time as an active dimension of reality, not merely a passive measure of change; the chemical atom as an expression of a generic idea of fundamental units with fixed properties, out of which nature as we experience it is composed; the ideas of the cell theory of life, the germ theory of disease, and the gene theory of inheritance, all conceptually allied to the atom idea; the ideas of energy, immaterial force fields, and structure and, thus, of relationships as elementary features of reality; the idea of systematically coupling science to technology, of coupling knowing to doing, and of using knowledge to synthesize a new world; the idea of evolution and its extension from biology to scientific thinking generally; and the idea that natural phenomena have a fundamentally probable and statistical character.

Lectures Twenty-Nine through Thirty-Five discuss central 20th-century scientific ideas, including the gene, relativity and quantum theories, the expanding Universe, computer science, information theory, molecular biology, and the idea of systems, especially self-organizing systems and the allied ideas of ecology and self-maintaining systems.

Appropriately, Lecture Thirty-Six concludes the course by reviewing the ideas that are distinctive of modern science and technology today and anticipating ideas likely to be drivers of change tomorrow, focusing in particular on cognitive neuroscience, biotechnology and nanotechnology, and physicists' search for a theory of everything.

Lecture One
Knowledge, Know-How, and Social Change

Scope: Science and science-based technologies became the primary drivers of social change by the late 19th century, broadening and deepening the impact of the first phase of the Industrial Revolution. Scientific ideas affect society primarily by way of technological innovations and secondarily through changing how we think of our selves and the world. Of all the scientific ideas that have shaped modern life, none is more influential than the idea of science itself in the form it was given by the 17th-century founders of modern science, a form in which the coordinate idea of techno-science was latent. Central to the idea of science is a conception of knowledge, formulated by the ancient Greek philosophers, as distinct from, and superior to, know-how. Ironically, increasingly sophisticated technological know-how long preceded the idea of science and continued to lead even modern science until the mid-19th century.

Outline

I. Science has changed our lives, but the questions of how it does so and why it is able to do so tell us as much about ourselves as they do about science.

  A. Beginning around 1800, science-linked technological innovation—*techno-science* for short—became the primary agent of social change, initially in the West, then globally.
   1. It is through technological innovation that science most directly affects how we live, physically and socially.
   2. With the Industrial Revolution—integrating the factory system of manufacture; mass-production machinery; and water, steam, and (in the late 19th century) electric power—an unprecedented and still-accelerating rate of innovation became *the* driving force of change in modern life.

  B. It is the ideas and discoveries of *modern* science that have changed our lives.

1. With very few exceptions, it is scientific *ideas* that affect us, not discoveries, which invariably turn out to be dependent on ideas for their understanding.
2. Modern science is that form of the study of nature that emerged in the 17th-century Scientific Revolution.
3. Modern science is an invention, a uniquely Western achievement, emerging only in the Christian culture of Western Europe.
4. Modern science is nevertheless deeply indebted to ancient Greek, Graeco-Roman, and Islamic sources; secondarily, to Chinese and Indian influences.
5. Although some scientific ideas have had a direct impact on how humans think of themselves, the world, and their place in the world, the greatest impact of science has been through techno-science.
6. Although the idea is Graeco-Roman, techno-science erupted into an agent of social change in the course of the 19th-century Industrial Revolution.
7. These lectures will demonstrate the assertion of the historian Lynn White that ideas and innovations only "open doors" for a society; they do not force a society to pass through those doors.
8. How a society responds to ideas and innovations is a function of values prevalent in that society.

II. This course offers a selective survey of major scientific ideas that have shaped our personal, social, and physical existence.
   A. It begins with the most influential of all scientific ideas, namely, the idea of science itself.
      1. This idea was an invention of ancient Greek philosophers that took on a decisive new form in the 17th century, the form we call modern science.
      2. It is from the idea of science, how science is conceptualized, that particular scientific ideas—theories of matter and energy, for example, of germs and genes, of cosmology and information—derive their force.
      3. Initially, our methodology will be to "reverse-engineer" the idea of science, exposing its key features and where they came from and asking why the idea of science was able to become a driver of social change via techno-science.

        4. The same ideas and innovations have different impacts on different societies; thus, these impacts give us valuable insights into societies and their values.
- **B.** This survey will be broadly chronological but not a systematic history, either of science or of individual scientific ideas.
    1. Each lecture will be self-contained, aimed at highlighting a single idea or development in a provocative way.
    2. But the lectures will be intensively cross-referenced in the way that the pieces of a mosaic image refer to one another.
    3. At the end, we will recover an integrated "picture" of science as a source of life- and society-changing ideas, revealing that science is not "natural" and that its social impact is not inevitable.
    4. The first six lectures unpack the idea of science, from the Sumerian invention of writing to the Graeco-Roman invention of the idea of techno-science.
    5. The second six explore the transmission of these ideas to modern Europe, from the 12th-century invention of the university to Copernicus's "revolutionary" theory of a moving Earth.
    6. Lectures Thirteen through Twenty-Eight address specific ideas and theories of modern science, from Francis Bacon and René Descartes on scientific method in the early 17th century to evolution and genetics in the late 19th century. These lectures call attention to a tension between two different conceptions of nature, one "atomistic" and the other process-based, and to the rise of life-transforming techno-science.
    7. Lectures Twenty-Nine through Thirty-Six discuss 20th-century theories that continue to shape our lives—quantum and relativity theories, cosmological theories and the ideas underlying computer technologies, information and systems theory, and molecular biology—and those theories likely to do so in the early 21st century.
- **III.** To appreciate that the idea of science inherited from the Greeks was an invention, we need to appreciate the truly astonishing amount of know-how humans accumulated without writing and without the Greek idea of knowledge.

A. From about 9000 B.C.E. to the onset of recorded history around 3000 B.C.E., humans became increasingly adept at increasingly complex technologies.
   1. They learned to domesticate plants and animals by way of highly selective breeding to create grains, fruits, and animals with specific characteristics.
   2. They invented and mastered increasingly sophisticated textile, ceramics, and metals technologies, including the mining and working of copper, bronze, iron, glass, gold, silver, lead, tin, and gemstones, as well as transportation and construction technologies, from boats and wheeled vehicles to cluster housing, irrigation canals and dams, fortifications, and monumental structures.
   3. Concurrently, people were living in increasingly large, typically fortified settlements and engaging in long-distance trade, which implies the creation of appropriate social institutions and social management "technologies."
   4. The earliest surviving written documents reflect the existence of long-established legal and moral norms, as well as commercial, social, and religious values and teachings.

B. We can readily infer, from the accumulation of know-how manifested by these prehistoric practices, a highly developed, probably implicit conception of what could be called knowledge.
   1. First of all, people could be classified as knowing how to do X or not knowing how to do X, thus as possessing or lacking knowing-how "knowledge."
   2. Of those who could be said to know how to do X, it was a matter of routine to distinguish those who were better at doing X from those who did X less well; that is, it was obvious how to rank the possession of knowing-how knowledge and without an absolute scale or standard!
   3. It was also obvious that some people did X creatively or innovatively, while others at most did X well in the traditional way.
   4. The fact that knowing-how knowledge was invented by our "primitive" ancestors and cumulated over millennia without written records is important to keep in mind.

5. Technologies *are* knowledge; they are, metaphorically speaking, "texts" that practitioners can "read," alter, and disseminate without writing.

Essential Reading:

Elizabeth Wayland Barber, *The Mummies of Urumchi*.

James E. McLellan III and Harold Dorn, *Science and Technology in World History*.

Questions to Consider:

1. Technologies have influenced social development for millennia, but what allowed technology to become a relentless driver of continuous social change in modern Western societies?
2. What can we infer about human beings from their artifacts in the 5,000 years before the first written records?

Lecture Two
Writing Makes Science Possible

Scope: Writing is a necessary but not a sufficient condition for modern science, for the kind of knowledge of nature that, coupled to technological innovation, is life-transforming. Modern science is wed to textuality, a legacy directly of the Renaissance embrace of printing and indirectly of the source of the idea of science in Greek philosophy, transmitted to the modern era via the medieval university. From the birth of modern science in the 17th century, it was a given that claims to knowledge of nature must be formulated in writing and disseminated via the written word: books, essays, articles, reports. The invention of writing in the 4th millennium B.C.E. in Sumer is the expression of an idea, coming after millennia of increasingly complex social interaction. It entailed the creation of a system of signs that evolved from idea-pictures to an alphabet and initiated a line of influence that, via Greece and Rome, links Sumerian cuneiform inscriptions on clay tablets to Internet-disseminated scientific journals.

Outline

I. Working backwards from the rise of modern science in the 17th century, writing appears as a necessary though not a sufficient condition for science.

   A. The idea of science as a formalized knowledge of nature is only known to us to have developed in literate cultures, and modern science emerged only in the "print-drunk" culture of Christian Western Europe.
      1. The invention of writing thus appears, at least empirically, to be a necessary condition both for the generic idea of science and for the specific idea of modern science.
      2. Writing is not a sufficient condition for either of these ideas, given that the former does not appear in all literate cultures and the latter did not emerge even in text-intensive Islamic, Chinese, or Indian cultures.

   B. *Science* is a name for knowledge defined in a particular way.

1. We saw in the previous lecture that know-how cumulated over millennia without writing.
2. Writing, therefore, is not a necessary condition for the creation, dissemination, and transmission of know-how, or practical knowledge.
3. Know-how is concretely embodied in particular objects, processes, and techniques and can be evaluated directly.
4. The knowledge that is at the root of the ideas of science and of modern science, however, has as its object not concrete experience but an abstract, *unexperienced* "reality."
5. The carrier of *cumulative and evolving* abstract knowledge effectively *must be* the written word.

II. Writing first appears in the archaeological record in the late 4th millennium B.C.E.
   A. The earliest written documents found to date come from the southeastern region of the so-called Fertile Crescent.
      1. This region was ruled by the Sumerians, a non-Semitic people who moved into the region and established a network of cities, among them, Ur, Nippur, Susa, and Uruk.
      2. The Sumerians invented a written form of their language that was inscribed on clay tablets with a stylus.
      3. This way of writing is called *cuneiform*, but the type, or system, of writing was *logographic/ideographic*.
   B. Formal *systems* of writing were long preceded by standardized tokens and inscription symbols.
      1. There is evidence, also from the Middle East, for standardized clay objects whose shapes encoded meanings long before writing systems.
      2. Pictographic seals were in use in Sumer centuries before writing and, like Sumerian writing, spread throughout the Middle East.
      3. Simple inscriptions, probably encoding names and numbers, also were widespread before the invention of writing and for long after.
   C. A writing system, like any invention, is the physical expression of an antecedent idea.

1. There is no record of the individual whose original idea the Sumerian writing system was, nor do we know why, all of a sudden, the idea both occurred to someone and "caught on."
2. The Chinese invented writing much later than the Sumerians and probably independently, and writing was invented in the Americas still later, in the 1st millennium B.C.E., but it appeared in Egypt shortly after it appeared in Sumer.
3. It is important to recognize that, like language itself, a system of writing is a *system*, having a holistic character, and thus, is an expression of an idea.
4. The earliest writing systems were ideographic and some, notably Sumerian, but not all, evolved into alphabetic systems.

III. The Sumerian invention of writing was extremely influential, and it was directly connected to the invention of the idea of science in ancient Greece.

   A. Hundreds of thousands of clay tablets with cuneiform writing in the Sumerian language have survived, the overwhelming majority of a commercial character—contracts, inventories, and wills—but tens of thousands are political, religious, and literary.
   1. The Sumerian writing system was adopted by the Semitic Akkadians, who adapted it to the requirements of their totally different language en route to establishing the first Babylonian Empire.
   2. The extensive surviving Akkadian literature includes the highly sophisticated legal codes of Ur Nammu and Hammurabi, as well as religious epics, songs, poems, and mathematical and astronomical texts.
   3. Following a pattern that repeats itself right down to the present, the availability of this new language "technology" created a positive feedback loop that multiplied many-fold the behavior it enabled.

   B. Over the next 2000 years, the originally Sumerian invention of writing spread eastward and westward from Sumer, evolving from logographic/ideographic writing to syllabic writing systems to purely alphabetic writing.
   1. The first alphabetic writing system emerged by the 14th century B.C.E., either in Ugarit (a city-state on Syria's Mediterranean coast) or further south, among the Phoenicians

(in modern Lebanon), by people still using variants of Sumerian cuneiform.
   2. Variants of the 22-letter Phoenician alphabet (Ugaritic used 30 letters) or of an earlier alphabet of which Phoenician was itself a variant (perhaps Ugaritic) became the ancient Hebrew script, perhaps as early as 1300 B.C.E., and later became the Arabic language script.
   3. Meanwhile, the Phoenicians, master merchants of the Mediterranean, taught their alphabet to the then non-literate Greeks around 800 B.C.E. and, a little later, to the Etruscans. Around 500 B.C.E., the Etruscans taught the alphabet to the Latins, better known to us as the Romans.
   4. In short, the earliest conceptions (by ancient Greek thinkers) of the idea of science and of scientific and technological ideas found expression more than 2000 years ago and are available to us today thanks to the Sumerians!
   5. The Greek response to writing was extraordinary, with books on philosophy, law, poetry, and drama literally pouring out by around 500 B.C.E.
   6. The philosophical idea of knowledge that became the cornerstone of modern science was formulated in this context.
C. Writing, like any technology, is first of all an idea.
   1. The Sumerians invented a system of writing, and it was extraordinarily influential, but like many technologies, it was not unique.
   2. Writing was almost certainly invented independently by the Chinese and again in Central America; the independent origin of Egyptian hieroglyphics, which appear only a few centuries after cuneiform tablets, is less clear.
   3. We know nothing about who invented writing or why.
   4. What was the necessity that provoked the invention of writing as a response?
   5. The Greek response is an excellent illustration of how an innovation "opened a door" for a society that chose to rush through that door.
   6. The Greek response to writing also illustrates a recurring feature of certain innovations: They become more valuable the more widespread their adoption.

7. Nevertheless, the spread of writing was not without its critics, ironically including Socrates, who wrote nothing but who founded Western philosophy through the writings of his student Plato.
8. In his dialogue called *Phaedrus*, Plato has Socrates deliver an impassioned argument against writing.

Essential Reading:

William V. Harris, *Ancient Literacy*.

Samuel Noah Kramer, *Sumerian Mythology*.

Questions to consider:

1. Is writing merely recorded speaking, or does writing have a distinctive relationship to thought and, thus, a character of its own, different from the relationship of speech to thought?
2. Given that technological know-how grew progressively more complex for millennia before writing was invented, *could* science, too, have grown as an orally disseminated teaching?

Lecture Three
Inventing Reason and Knowledge

Scope: As evidenced by the archaeological record, people were reasoning effectively for millennia before the first Greek philosophers, learning how to do a great many complicated, "unnatural" things. Nevertheless, between 500 B.C.E. and 350 B.C.E., Greek philosophers who were subsequently highly influential argued for particular conceptions of reason, knowledge, truth, and reality. Their abstract and theoretical definition of knowledge—as universal, necessary, and certain—contrasted sharply with an empirical, concrete, and practical definition of knowledge. This contrast continues to this day in the distinction we make between science and engineering, between "true" understanding and "mere" know-how. One of the most lasting Greek cultural achievements was the invention of the discipline we call logic by codifying reasoning in a way that supported the philosophical definition of knowledge. The division of reasoning into deductive, inductive, and persuasive forms of argument played a fundamental role in the invention of the idea of science as knowledge of nature.

Outline

I. As an idea, knowledge had to be invented.
   A. Plato's Socrates opposed writing yet appears as a teacher in a large body of writings by Plato, through which Socrates teaches philosophy.
      1. Plato's dialogues and the still larger body of writings by his student Aristotle are the foundations of Western philosophy.
      2. A central issue for both Plato and Aristotle is the nature of knowledge.
      3. The question "What does the word *knowledge* mean?" does not have an obvious answer.
      4. There is no objective, absolute dictionary in which to look up the meanings of words: People define words!
   B. Recall that the accumulation of know-how is quite independent of a formal *idea* or explicit concept of knowledge.

1. Knowledge could have been defined in terms of know-how, that is, as effective practice.
2. Indeed, those Greek philosophers called Sophists defined knowledge in this way.
3. Plato and Aristotle, however, rejected such a characterization in favor of a highly abstract and intellectualized definition.
4. For them, knowledge was universal, necessary, and certain, hence, timeless.

C. In this abstract definition, knowledge is divorced from experience.
1. Experience produces know-how, but both are particular, context dependent, merely probable, and subject to change over time.
2. Philosophical knowledge has as its object a reality behind experience that is accessible only to the mind, which can only arrive at such knowledge by reasoning.
3. The paradigm of such reasoning for Plato is mathematics, which gives us universal, necessary, and certain knowledge and, thus, validates his definition.
4. Humans had accumulated powerful forms of know-how for millennia, yet the Platonic-Aristotelian definition of knowledge has dominated Western philosophy.
5. It was the genius of Greek philosophers to codify—that is, to organize in a systematic way—the activity of mind they called "reasoning" and to define certain products of that activity in ways that denied to know-how the status of knowledge and to informal/implicit reasoning the status of reason.
6. In the process, they created the disciplines we call logic and mathematics; the idea of a "proof" of a truth claim; and a cluster of correlated, abstract ideas of reason, knowledge, truth, and reality, out of which the idea of science and, much later, modern science itself emerged.

II. Aristotle codified the idea that reasoning meant drawing inferences—inferring the truth of one statement from the truth of others—and he organized reasoning into three distinct modes of inferring whose study makes up the discipline called logic: deduction, induction, and dialectic/rhetoric.

A. *Deductive inference* is the ideal mode of reasoning.

1. Deduction is ideal because, given the truth of the premises from which they are drawn, deductive inferences are necessarily true and, thus, certain, and they can be universal truths.
2. Especially in his books *Prior Analytics* and *Posterior Analytics*, Aristotle organized in a systematic (but highly selective) way the teachings of earlier Greek philosophers.
3. In this, he did for the study of reasoning what Euclid did for geometry, and like Euclid's *Elements*, Aristotle's collection of logic texts, called the *Organon*, was required reading in all Western universities through the 19th century.
4. Aristotle showed that the truth of the conclusion of a deductive argument—he focused on a particular type of argument called *syllogisms*—was strictly a matter of the form of the argument, not its content.
5. This makes the question of how we can know the truth of the premises of a deductive argument—at least one of which must be universal—a crucial one to answer.
6. The problem is that the particularity of sense experience implies that generalizations from experience can never be certain.
7. Aristotle explicitly links knowing the truth of the premises of deductive arguments to knowledge of nature (and, thus, to the possibility of science).
8. Note well the intimate connection here among the ideas of reasoning, truth, knowledge, and reality, especially the concept that *knowledge* means that whose truth is certain and that the object of knowledge and truth is not sense experience but the way things "really" are.

B. *Inductive reasoning* is not just second best for Aristotle; it's a totally different *kind* of inference-drawing from deduction.
1. Inductive inferences are always probable, never certain.
2. The truth of inductive inferences is *not* a matter of the form of inductive arguments: Content matters.
3. Aristotle describes inductive inference as reasoning from particulars to universals—but also from effects to causes—which is why content, derived from experience, matters and why certainty cannot be reached.

    4. Nevertheless, Aristotle assigns to induction a role in our knowledge of the truth of the universal statements about reality/nature that serve as premises in deductive arguments.
    5. Without such knowledge, we cannot have a science of nature, and the founders of modern science, especially Francis Bacon and René Descartes, had to solve the problem Aristotle posed.
    6. Keep this in mind for Lecture Thirteen: Experimentation cannot bridge the logical gulf between induction and deduction!
  C. *Dialectic* and a related form of arguing called *rhetoric* complete the science of logic.
    1. Strictly speaking, dialectic for Aristotle refers to arguments whose premises we *accept* as true, hypothetically, without *knowing* that they are true.
    2. Dialectical reasoning allows us to explore the deductive inferences that can be drawn from a set of premises *if* we accept them as true.
    3. The form of dialectical arguments is deductive; thus, although the conclusions of such arguments are necessarily true *logically*, they may not be true "*really*" because the premises may not, in fact, be true.
    4. Keep this form of reasoning in mind when we reach the 17th century and the idea of a scientific method that uses inductive-experimental reasoning to validate hypotheses that allow us to have universal and necessary knowledge of nature.
    5. Speaking less strictly, dialectic overlaps rhetoric, which is the art of persuasion, using arguments that may appear logical but, in fact, are not.
    6. Aristotle's book entitled *Rhetoric* deals in part with the kind of reasoning we use every day to reach action decisions, given that we never possess knowledge that would allow us to deduce what to do in a specific situation.

III. From the perspective of modern science, the relation between induction and deduction is especially important.
  A. The experimental method is inductive, but a science of nature that emulated mathematical reasoning would have to be deductive.
    1. Francis Bacon seemed to believe that careful experimentation could bridge induction and deduction.

      2. He was wrong about this because, logically speaking, induction and deduction cannot be bridged.
      3. Generalizations arrived at inductively are, in principle, only probably true, while the premises of deductive arguments must be true necessarily.
  B. For Descartes, as for Aristotle, science as knowledge of nature is based on deduction.
      1. Where are the premises or principles of nature to come from that will allow us to explain phenomena deductively?
      2. This is a recurring issue in science.
      3. Aristotle and Descartes decided that we knew the truth of some statements about nature and about reasoning intuitively, but this was a controversial position in modern science.

Essential Reading:

Aristotle, *Prior Analytics*, *Posterior Analytics*, and *Rhetoric*.

Questions to Consider:

1. Are there uniquely correct definitions of the words we use?
2. Why was the philosophical conception of knowledge triumphant given the manifestly superior practical value of defining knowledge as know-how?

Lecture Four
The Birth of Natural Science

Scope: For us, *science* simply means a particular approach to the study of natural and social phenomena and a body of knowledge generated by that approach that has evolved over the past 400 years. Initially, however, *science* meant universal, necessary, and certain knowledge generically. Even those Greek philosophers who adopted this definition of knowledge disagreed as to whether such knowledge *of nature* was possible, Plato arguing that it wasn't and Aristotle that it was. Greek philosophical theories of nature long predate Plato and Aristotle, but their ideas influenced Western culture most deeply, right into the 20th century. In addition to his codification of logic, Aristotle's theories in physics and biology and his underlying naturalistic metaphysics and empirical methodology dominated Western nature philosophy through the 16th century. Modern science defined itself in opposition to these theories even as Aristotle's ideas and logic continued to inform modern science.

Outline

I. Plato formulated a rigorous, generic idea of knowledge that was tied to logic and mathematics, but it was Aristotle who formulated the specific idea of knowledge of *nature*, which is what we typically mean by *science*.

    A. The relation between mathematics-based knowledge and knowledge of nature is not self-evident.

        1. Mathematics is unquestionably knowledge in the full philosophical sense, but it is not clear what it is knowledge *of*.

        2. Mathematics may give us knowledge of objects that are created by the mind, in the manner of a game, or of objects that exist independently of the mind.

        3. If the latter, these objects may be natural and part of experience or supra-natural and accessible only in the mind.

        4. Knowledge of nature, on the other hand, must be about what is independent of the mind and part of experience.

5. Mathematics works when applied to experience, so it's not just a game, but it seems impossible to derive mathematical knowledge from experience.

B. It was Aristotle, not Plato, who formulated the idea of knowledge of nature, and he formulated as well the single most influential theory of nature in Western cultural history.
1. Aristotle was a "scientist" as well as a philosopher.
2. The 17th-century Scientific Revolution was, in large part, a "revolt" against his theory of nature and his scientific ideas.

C. Aristotle created a comprehensive theory of knowledge of nature.
1. This theory was grounded in Plato's definition of knowledge as universal, necessary, and certain, and it explained how the human mind could have knowledge of nature given that definition.
2. The idea of science, as we understand that term, is thus, first of all, indebted to a particular idea of knowledge adopted by Aristotle allied to Aristotle's idea of nature.
3. Aristotle argued that the logical gulf between induction and deduction could not be bridged by sensory experience, but it could be bridged by the mind.

D. With knowledge of nature established as a possibility, Aristotle embedded his theory of nature in a metaphysics, that is, in a set of absolute, timeless, universal principles that define what is real.
1. One of these principles is that nature is all that there is, that the real *is* the natural.
2. Plato had argued (and perhaps believed) that the real was primarily ideal—his utopian realm of immaterial, universal forms—and that the natural, the realm of form allied to matter, was inferior to, and dependent upon, the ideal.
3. It followed that, for Plato, knowledge of nature was not possible because the natural was particular and continually changing.
4. For Plato, reality was accessible only to the mind and definitely not through the senses-based experiences of the body.
5. Aristotle accepted that form and matter were the ultimate categories of reality, but another of his metaphysical principles was that everything real was a *combination* of form

and matter; thus, the real and the natural were one and the same.
6. Subsequently, *this* idea of Aristotle's became a fundamental principle of modern science, namely, that in studying nature we are studying reality: that there is nothing real that is not natural.
7. Another way of putting this principle is that nature is a self-contained system, and we will see that precisely this claim was central to the medieval revival of Aristotle's theory of nature and to modern science.
8. Note, too, how Aristotle's metaphysical principles are analogous to the universal laws of nature proposed by modern science.

II. Knowledge, Plato and Aristotle agree, is universal and timeless, while nature is particular and continually changing; thus, Plato seems right in concluding that science is impossible.
  A. Aristotle's theory of knowledge explains how the mind abstracts universals from experience, and his theory of nature explains how we can have knowledge of change.
    1. He postulated the existence of a mental faculty, the active intellect, that could recognize in individual objects the intrinsic universal forms of which particular natural objects and phenomena were instances, for example, recognizing in a particular dog such universals as species, genus, order, and class.
    2. Knowing "intuitively" the truth of universals, the mind can then achieve knowledge by creating deductive accounts of nature.
    3. That is, Aristotle proposed bridging experience-based induction and deductive knowledge by intuitive knowledge of universals that become the premises of deductive arguments.
  B. Plato and Aristotle on knowledge of nature illustrate a recurring feature of intellectual history.
    1. The history of ideas, like human history generally, is rarely linear and logical.
    2. Plato borrowed selectively from his predecessors in formulating his ideas about form, matter, knowledge, and mathematics.

3. Aristotle borrowed selectively from his teacher Plato and from those same predecessors, but his selection was different from Plato's.
4. In particular, Plato adopted certain Pythagorean ideas about mathematics but rejected the idea that mathematical objects existed within the natural.
5. Aristotle rejected the idea that mathematics was central to knowledge of nature, except for optics, music, and astronomy, which he considered special cases.

C. Aristotle's theory of nature is dominated by a theory of change, and his physics, by a theory of motion as one type of change.
   1. Aristotle's theory of change begins with his famous "four causes" analysis of change.
   2. Change is explained when it is related to four causes, a term that is suggestive of reasons for, or principles or parameters of, change but corresponds only loosely to what *cause* came to mean in modern science.
   3. Note the absence of predictive success and control of experience as criteria of knowledge of nature, criteria considered important in modern science.
   4. For Aristotle, as for Plato, knowledge is abstract and theoretical *only*.
   5. Aristotle's four causes are material, formal, efficient, and final.
   6. Aristotle's theory of nature, including his physics, was thus qualitative because he held that, except for astronomy, optics, and music, mathematics was only incidentally relevant to explaining natural phenomena.

III. Aristotle's physics, though enormously influential, does not begin to exhaust the scientific ideas invented by ancient Greek thinkers that continue to affect our lives today through their incorporation into modern science. Several of these ideas are highlighted below.

A. The idea that everything that is forms a *cosmos*, an ordered whole.
   1. An ordered whole implies a structure, a system of relationships.
   2. This makes the task of cosmology explaining that structure.

B. The Pythagorean idea that mathematics is the basis of all knowledge of nature.

        1. In the 16th and 17th centuries, this idea was attributed to Plato.
        2. One of the defining characteristics of modern science is its mathematical character, contrary to Aristotle's view.
    C. The idea that reality is composed of timeless, elementary substances with fixed properties.
        1. This *substance metaphysics* derives from the writings of Parmenides, cryptic even in antiquity.
        2. It was the inspiration for postulating elementary atoms, whose combinations, based on their innate, fixed properties, are the cause of all complex things and all phenomena.
    D. The rival idea that there are no elementary substances because reality is ultimately a web of rule-governed processes.
        1. The philosopher Heraclitus promoted this *process metaphysics* against Parmenides and the atomists.
        2. Since the mid-19th century, this idea has become increasingly prominent in science.
    E. The idea that nature is ultimately matter in motion and no more than that.
        1. Aristotle's idea that nature *is* reality reduces to materialism if his notion of form is interpreted as a pattern of motion.
        2. This is exactly what the founders of modern science did.

Essential Reading:

Aristotle, *Physics*; also *On Generation and Corruption, Generation of Animals*, and *On the Soul*.

Lucretius, *On the Nature of the Universe*.

Questions to Consider:

1. How is it that without any instruments and with the most limited experience and experimentation, Greek philosophers formulated so many concepts and ideas that continue to be central to modern science?
2. What is missing from an Aristotelian causal explanation that we expect from science?

Lecture Five
Mathematics as the Order of Nature

Scope: Together with the experimental method and deductive reasoning, the use of mathematics is the hallmark of modern science. Mathematics is the language of scientific explanation because it is, in some sense, the "language" of nature. That mathematical forms are the essence of natural objects and processes was central to the teachings of Pythagoras and his followers. The orderliness of natural phenomena is a mathematical order, as in the difference between music and noise. Beginning with Pythagoras, Greek mathematics, based on geometry primarily, became the paradigm of using deductive reasoning to acquire philosophical knowledge. Archimedes used deduction and mathematics to solve problems in physics. Aristotle dismissed the value of mathematics for natural science, except for music, optics, and astronomy, but Renaissance philosophers, partly under the influence of Plato's writings, restored mathematics to its Pythagorean position.

Outline

I. Looking back from the present, perhaps the single most influential Greek scientific idea was the idea that nature is, in its essence, mathematical.
  - **A.** Quantifying nature was one of the core distinctive features of modern science in the 17th century.
    1. This was obviously the case for physics then, and quantification in all of the sciences has become progressively more intense ever since.
    2. Increasingly, the most advanced scientific instruments—for example, particle accelerators—generate numbers that are interpreted as features of nature or converted into "pictures" by computer programs.
  - **B.** From the perspective of modern science, the mathematization of nature began with Pythagoras, literally or mythically.
    1. Pythagoras certainly was a real person, who flourished circa 500 B.C.E. and taught that *number* was the basis of the orderliness of experience.

2. He founded a mystical religious-philosophical movement keyed to this teaching that survived for centuries, with communes and schools across the Greek cultural world.
   3. For Pythagoras, mathematical forms were physically real, not mere abstractions.
   4. There was a great deal of specific number know-how before Pythagoras, as revealed by surviving Babylonian and Egyptian texts, but not mathematics as a distinctive body of knowledge.
  C. The idea of a deductive proof is what distinguishes what we mean by mathematics from number know-how.
   1. Pythagoras invented the idea of a general proof of the relationship between the sides and the hypotenuse of a right triangle, in the process, transforming number know-how into mathematics.
   2. The idea of the proof is connected to the codification of deductive reasoning as the defining form of reasoning for mathematics, as it has been ever since the time of Pythagoras.
   3. Deductive reasoning, because it generated universal, necessary, and certain truth in mathematics, became the defining form of truth-generating reasoning for knowledge generally and, thus, for science as knowledge of nature, as well.

II. Pythagoras invented two great scientific ideas: the idea of a logical proof of general truths about quantitative relationships and the idea that quantitative relationships were the essence of the physical world.
  A. Pythagoras, supposedly stimulated by an actual experience, concluded that the difference between music and noise, for example, was mathematical proportion.
   1. Pythagoras invented the monochord—a single taut string with a movable fret tuned to a given note—to demonstrate this "discovery."
   2. He showed that all the notes of the octave could be generated by moving the fret precise quantitative distances.
   3. If the initial note sounded by plucking the string was a middle C, for example, then halving the length of the plucked string sounded C an octave higher, while doubling the length sounded C an octave lower (keeping the tension constant throughout).

B. Pythagoras's teaching that mathematical forms were physically real, as illustrated by music, entails a mathematical metaphysics.
   1. That is, it entails believing that mathematics, now conceived as rigorously logical truths about numbers and numerical relationships, is the ultimate reality.
   2. Mathematical form is the indwelling orderliness of the physical world that we experience.
   3. But this poses profound problems when it can be proven that there are mathematically incoherent aspects of the world.
   4. Using the proof technique he invented, Pythagoras or his followers proved that certain mathematical forms that were physically real were irrational: could not be expressed as ratios of whole numbers.
   5. Later, the need for "imaginary" numbers to solve physically real problems posed a similar disquiet.
   6. The implication was that, ultimately, the world was not rational, that the orderliness of nature was superficial!

C. The power and influence of Pythagoras's coupling of mathematical form and physical reality are reflected in the history of music.
   1. His insight that whole numerical ratios are what distinguish music from noise became the basis for millennia of passionate disputes over musical tuning systems.
   2. As with the diagonal of a square, it turns out that the ratios that generate the notes of the octave "break down" when extended up and down the musical scale.
   3. They generate dissonances, which are noise, not music!
   4. These controversies, already lively in the ancient world, led to multiple mathematical schemes for tuning instruments.
   5. Pythagoras's insight that mathematical form was the essence of music was preserved, but his tuning system was not.

D. These controversies became truly vicious, however, when Western musical harmonies became more complex during the Renaissance.
   1. Vincenzo Galilei, Galileo's father, was an active participant in these controversies, which involved sophisticated mathematics, skill in musical performance and composition, and innovative experimentation.
   2. The musical stakes were high because the tuning system employed affected instrument design, composition (which

combinations of notes would sound harmonious), and of course, performance.
   3. The scientific stakes were even higher because of the increasing identification of mathematics with physics during the Renaissance and in early modern science.
E. From the beginning, Pythagoras used the idea that nature was mathematical to solve scientific problems, and this became central to modern science.
   1. Descartes identified geometry with space, as did Einstein in his general theory of relativity, though in a very different way.
   2. Galileo called mathematics the "language" of nature, stating that this was a Platonic idea though he knew it was Pythagorean.
   3. Galileo surely wanted to avoid the association with mysticism and magic that had grown up around Pythagorean teachings.
   4. It was, however, misleading to call his view Platonic because Plato denied that knowledge of nature was possible!
F. Pythagoras himself used his mathematical metaphysics to set an agenda for Western astronomy that was accepted for more than 2000 years.
   1. He taught that the planets, stars, Moon, and Earth were spheres because that was a "perfect" mathematical shape.
   2. That the Earth was a sphere was fundamental to all Western astronomy: Aristotle offered three different proofs, and it was taught in all the universities centuries before Columbus.
   3. Pythagoras also concluded, on grounds of mathematical "perfection," that the form of the orbits of the planets, Moon, Sun, and stars around the Earth was a circular path and that they moved at uniform speeds.
   4. Observationally, this seems not to be true, yet for more than 2000 years—up to and including the great Galileo!—Western and Islamic astronomers made it the goal of their theories to explain how the apparent motions were generated by "real" motions that satisfied Pythagoras's criteria.
   5. This was the basis of Ptolemy's ingenious, highly contorted, epicyclic Earth-centered theory of the heavens that dominated Western astronomy until Copernicus's moving-Earth theory initiated modern astronomical science.

      6. It was not Copernicus, however, but Johannes Kepler who first rejected Pythagoras's entire agenda, arguing in 1609 that the planets moved in ellipses and, in 1619, that planets moved at non-uniform speeds.

G. Returning to earlier history, in the 3rd century B.C.E., Pythagoras's idea for a mathematical physics was developed further by Archimedes.

      1. Archimedes, killed by a Roman soldier in 212 B.C.E., exemplifies what became the role model for mathematical physics in the modern period.

      2. Archimedes was a great mathematician, extending geometry, pioneering trigonometry, and inventing a precursor of the Newton-Leibniz calculus.

      3. Concurrently, he applied mathematics to physical and even technological problems—for example, in hydrostatics and the theory and design of machines—while insisting on maintaining mathematico-deductive reasoning.

      4. With the reprinting of texts by Archimedes in the 16th century, the Pythagorean-Archimedean idea of mathematical physics was adopted by Galileo and, through him, became central to the invention of modern science.

Essential Reading:

Euclid, *Euclid's Elements*.

Jacob Klein, *Greek Mathematical Thought and the Origin of Algebra*.

Questions to Consider:

1. If mathematics derives from experience, how can it be universal, necessary, and certain? But if it does not derive from experience, how can we know it?

2. Mathematics "works" in science, as evidenced by the predictive success of theories and their power to control experience, but why does it work?

Lecture Six
The Birth of Techno-Science

Scope: There is a qualitative difference between the impact of technology on society from the Sumerian invention of writing to the 19th century and its impact since then. The difference largely reflects 19th-century innovations that exploited scientific knowledge and institutions created to stimulate this "techno-science" and channel its products into society. Techno-science became a powerful driver of change in the 19th century, but the idea that the best practice/know-how is based on theory/knowledge first appears in the Graeco-Roman period. Archimedes not only pioneered mathematical physics, but he also developed a mathematics-based theory of machines and mechanical devices. The early-20th-century discovery of a complex machine dating from this period revealed that Greeks built machines to express theories. Vitruvius, in his influential book *On Architecture*, promoted the idea that effective engineering must be grounded in science, and Ptolemy used sophisticated mathematics to generate accurate maps of the heavens and the Earth.

Outline

I. In the Roman Empire, practical know-how/craftsmanship began a transition to knowledge-based engineering—techno-science—that would be resumed in the Renaissance and erupt in the 19th century into the leading agent of social change.

   A. The first steps toward late-19th-century techno-science were taken in the late years of the Roman Republic and the first centuries of the empire, roughly 200 B.C.E. to 300 C.E.
   1. Rome conquered Greece in the mid-2nd century B.C.E. and assimilated the continuing practice of Greek philosophy, science, mathematics, and technology into Roman cultural and social life.
   2. By that time, the idea of science as mathematics-based, abstract/deductive knowledge of nature had crystallized out of Greek philosophical thought.

      3. In addition, the first steps toward techno-science had been taken by Greek inventors and mathematicians.
- **B.** A pivotal figure in the dissemination of the idea of techno-science was the 1st-century-B.C.E. Roman "architect" Vitruvius.
  1. The term *architect* then was closer to what we mean today by *engineer*.
  2. An architect was, in effect, a civil-cum-mechanical engineer who was expected to design the devices or constructions whose realization he then oversaw.
  3. Vitruvius was influential in his own time but far more influential in the Renaissance and early modern period.
- **C.** Vitruvius wrote an "engineering" handbook that was rediscovered in the very late 15th century and repeatedly reprinted over the next 200 years, influencing the rise of modern engineering.
  1. His book *On Architecture* is effectively a handbook for engineers.
  2. Its chapters cover siting cities; laying out streets; and designing and constructing roads, houses, water systems, and public buildings.
  3. The public buildings include theaters, and Vitruvius refers to techniques of "naturalistic" painting for scenic backdrops.
  4. The ninth chapter of the book is about time-keeping and the tenth is primarily about the range of machines available to architects to do the civil or military work they need to do.
  5. Both of these sections and his distinction of "engines" from machines illustrate perfectly Vitruvius's claim in the opening paragraph of *On Architecture* that the architect must be "equipped with knowledge" of a kind that is the "child of practice and theory."
- **D.** Vitruvius's claim that that the architect/engineer "must be equipped with knowledge" and with knowledge of many fields is a true intellectual innovation.
  1. For Plato and Aristotle, knowledge is fundamentally different from, and superior to, know-how.
  2. For Vitruvius to simply state as if it were self-evident that the best know-how is linked to knowledge is startling in context.
  3. Knowledge yields understanding and is the "pure" pursuit of truth, while know-how is the "impure" pursuit of effective action in some particular context.

      4. This distinction between theory and practice surfaced as a major source of tension in the 19th century, for example, in attempts in the United States to couple science to technology.
      5. Vitruvius was not the first to *apply* abstract knowledge to concrete know-how, but he may have been the first to formulate the idea that know-how *should be* based on knowledge as a matter of principle.
      6. This marks the birth of the idea of techno-science.
  E. Vitruvius's claim that the union of theory and practice is fertile is an important one.
      1. Theory seeks to explain by deductive "demonstration," exemplified by geometry, the favored form of mathematics for the Greeks, as it was even for Galileo and Newton in the 17th century.
      2. Practice, by contrast, devises what works and uses it whether or not the practitioner can explain why it works.
      3. Vitruvius acknowledged Greek predecessors from the 3rd century B.C.E. on who applied knowledge to create new kinds of machines.
      4. The claim that the effective design and use of machines and engines depends on knowledge from which design and use can be deduced is, however, an innovation in its own right.
      5. In the process of making his claim, Vitruvius also created a "space" for a new figure, which we recognize as an engineer, who stands between the scientist and the craftsman and possesses a distinctive expertise of his own.
      6. As we will see, this engineer comes into his own in the 19th century and plays a key role in uniting ideas and innovations.

II. Vitruvius explains that he has learned what he knows from reading in the works of his great predecessors.
  A. He names dozens of authors of books that he has studied on the subjects that he writes about in his book.
      1. Among these are Ctesibius of Alexandria, Archimedes, and Philo (Philon) of Byzantium, all from the 3rd century B.C.E.
      2. Ctesibius invented a family of machines based on force-pumped water and air, as well as a complex water clock and improved military catapults. Although his book disappeared, his machines were influential for centuries.

3. We already know Archimedes as a great mathematician and founder of mathematical physics, but he also was noted for contributions to applied physics, especially his theoretical account of the five basic machine types, and for civil and military applications of machines.
4. In antiquity, Archimedes was said to have built an orrery, but this was discounted as far too complex a machine for him to have built until the discovery and analysis of the so-called *Antikythera machine*.
5. Dredged up early in the 20th century off the coast of the Greek island of Antikythera, this device, almost certainly from the 2nd century B.C.E., is easily the most sophisticated machine that would be made in Europe for at least 1500 years.
6. X-ray analysis reveals that the Antikythera machine is an extraordinary combination of precisely machined, interconnected gears that probably made complex calendrical and/or astronomical calculations, then displayed them on its "face" (now decayed).
7. This artifact, whatever its precise purpose, is clear evidence of a machine designed and constructed to embody ideas and theoretical knowledge.
8. It is a machine that required theoretical knowledge for its conception and design, together with carefully coordinated specialized know-how for its construction.

B. Vitruvius also stands as one figure in a long, innovative Roman tradition of sophisticated architecture/engineering.
1. On the one hand, there are the machine designers, such as Hero (Heron) of Alexandria. Hero echoes Vitruvius's union of theory and practice, science and engineering, and describes dozens of possible machines using steam, water, and air power.
2. On the other hand, there is the astonishingly sophisticated technological infrastructure of imperial Rome.
3. The water supply system and the network of public baths in the city of Rome, serving a million people, still inspire awe, especially the Baths of Caracalla and those of Diocletian.
4. The ruins of Pompeii attest to the existence of a central water supply system metered at each home.

5. The Romans both could and did build water-powered grain mills but on a limited scale.
6. For Vitruvius, Archimedes exemplifies the linkage between theory and practice, knowledge and know-how that distinguishes the architect/engineer from lesser craftsmen.
7. More generally, we need to appreciate the existence of a rich and technologically innovative intellectual tradition in Greek and Graeco-Roman mechanics, both before and after Vitruvius, a tradition that was imported wholesale into Islamic culture from the 8th century on.

Essential Reading:

A. G. Drachmann, *The Mechanical Technology of Greek and Roman Antiquity*.

Vitruvius, *The Ten Books on Architecture*.

Questions to Consider:

1. Given the weakness of theory in relation to practice in antiquity, why would anyone claim that, ideally, practice must be based on theory?
2. Why didn't the Greeks and Romans build calculators and water- and steam-powered machines that they seem to have known how to build?

Lecture Seven
Universities Relaunch the Idea of Knowledge

Scope: Schooling is not unique to a text-based culture, but it is necessary, because writing must be taught and, as Socrates observed, because texts must be explained, clarified, and interpreted. The university was invented in the 12th century as the "ultimate" school: not merely a higher-level school but a community of scholars pursuing knowledge for its own sake. The text-based intellectual culture of Greece and Rome and, subsequently, of Islam made the medieval university inevitable once the desire to possess that culture arose. It made the Greek idea of science, in both the generic and specific senses, central to western European cultural life. From the 12th through the 16th centuries, universities created a population of educated people exposed to Classical, Islamic, and contemporary mathematical, philosophical, medical, and scientific texts. Universities thereby made available to the founders of modern science intellectual "tools" they used to create modern science.

Outline

I. The university in the Christian West was an innovation with profound intellectual and social consequences.
   A. The collapse of the Roman Empire in the 5th century marked the onset of an intellectual "Dark Age" in western Europe.
      1. The period from roughly 400–1000 was a time of significant physical, social, and political change in western Europe, but virtually nothing innovative was done in philosophy, science, mathematics, or technology.
      2. This began to change in the 11th century, and by the 12th century, there was a sudden explosion of cultural, social, commercial, and intellectual dynamism.
      3. This amounted to a cultural "renascence" that differed from the better known Renaissance in not seeking to revive the culture of antiquity.
   B. There had been schools of one kind and another in Europe, some of them advanced, at least since the 5th century B.C.E.

1. In fact, Babylonian cuneiform tablets and Egyptian papyri from the 2nd millennium B.C.E. are clearly textbooks for teaching arithmetic, geometry, and medicine.
2. Hippocrates's medical school on the island of Kos and a rival school in Cnidus date to the 5th century B.C.E.
3. Hippocrates transformed the healing "art" by using texts to transmit ideas along with information, leading to the organization of this information into knowledge by way of medical theories.
4. The followers of Pythagoras maintained a network of schools from the 4th century B.C.E., and Plato and Aristotle founded formal schools of philosophy in Athens that remained open until the 6th century.
5. From the 9th century, there were Islamic colleges of advanced learning that were known to Europeans, especially those that taught medicine, and analogues in China and India.

C. The Catholic Church maintained a hierarchical school system throughout the so-called Dark Age.
1. There was a need for priests, of course, but also for civil servants, clerks, and administrators for the Church and the secular government, as well as for doctors and lawyers.
2. Local monastic schools fed regional cathedral schools, and beginning in the 10th century, there were medical and law schools in Italy and France.
3. In the 12th century, there was an enormous increase in the student populations at the cathedral schools of Paris and Bologna especially, leading to the creation of the first universities there.
4. The university idea spread rapidly, especially in western and southern Europe, but with dozens of universities created by 1400, from Oxford and Cambridge in England to Krakow in Poland.

II. The medieval university was a new kind of social-intellectual institution.
A. The four recognized faculties/degrees were law, medicine, theology, and philosophy, with philosophy encompassing all secular learning other than medicine and law.

1. This would remain the case until the 19th century, when philosophy fragmented into modern disciplines ranging from the humanities to the physical sciences and mathematics.
2. The invention of the university, like the invention of writing, emerges as a major scientific idea from the reverse engineering of modern science.
3. The pursuit of secular knowledge in the philosophy and medical faculties became an essential condition for the emergence of modern science in the 17th century and for the maturation of modern science into a driver of social change in the 19th and 20th centuries.

B. The key to appreciating the university as a social-intellectual innovation is the value attributed to secular knowledge for its own sake.
1. The Church immediately recognized the potential threat of the universities to religious teachings and values but tolerated them nonetheless.
2. The size of the university population by 1400 constituted a new social and commercial phenomenon.
3. The collateral growth of secular governments claiming independence from the Church was closely tied to the university movement.
4. Socially, the degree-granting structure echoed the familiar craft guild structure.

C. The secular dimension of the university was not at all a rebellion against religion or the religious establishment.
1. The primary assertion was the legitimacy of the pursuit of secular knowledge for its own sake (not for the sake of salvation) and the autonomy of that pursuit.
2. Both of these claims were considered to be consistent with the faith-based character of medieval European society.
3. The explicit tension between these claims and the claims of faith is manifested in the struggle between the philosopher Peter Abelard and Bernard of Clairvaux, one of the leading churchmen of the 12th century.

D. Abelard effectively resurrected the naturalist philosophy of Aristotle, whose writings were newly available in Latin in the 12th century.

1. This philosophy incorporated Aristotle's idea of knowledge as a body of universal, necessary, and certain truths generated by deductive reasoning.
2. Aristotle's intellectualism and naturalism were reflected in Abelard's claim that the human mind could arrive at such knowledge on its own and that humans could achieve the good on their own.
3. Bernard of Clairvaux rejected the possibility of deducing religious truths independent of faith and charged Abelard with heresy for teaching that humans could be saved by their own efforts.
4. Bernard destroyed Abelard, but even he did not crush the newborn universities when he could have done so.

III. The university thus emerges as a secular social phenomenon in the 12th century, an institution sheltering the study of secular knowledge within a faith-based community and loyal to its religious teachings and values.

A. The university must be understood as a *response* to a social demand for such an institution, not as *creating* that demand.
1. Once created, however, and as an institution dedicated to the text-based study of knowledge, which texts were to be studied?
2. Overwhelmingly, the texts came from ancient Greek and Roman writers on mathematics, philosophy, astronomy, medicine, and law.
3. Beginning in the 12th century, contemporary thinkers, such as Abelard, were generating important new texts.
4. Concurrently, and quite independently, the assertion of the legitimacy and power of Aristotelian reasoning within a faith-based community was made in Islam and Judaism.
5. Averroës, the greatest Islamic philosopher of the period, and Maimonides, the greatest Jewish philosopher, both championed an Abelard-like intellectualism while proclaiming their religious orthodoxy.
6. Nevertheless, like Abelard, each was severely criticized by defenders of the absoluteness of religious truth, and Averroës came close to execution as a heretic, yet both became important influences on late-medieval Christian philosophy!

B. The university assimilated the Platonic-Aristotelian definition of knowledge for all the subject areas studied under the rubric of "philosophy," including nature.
   1. Only such a definition could put knowledge on a par with revelation as a source of truth.
   2. Inevitably, therefore, philosophy encountered the problem of knowing the truth of universal statements that could serve as the premises of deductive arguments.
   3. A range of mutually exclusive solutions to this problem was proposed, but the problem endures within science even today.
   4. The philosophical study of nature in the university generated a set of ideas that was subsequently incorporated into modern science, which ties the idea of the university more closely to the idea of science.
   5. These ideas included: that nature was a closed system, that explanations of natural phenomena could not invoke supra-natural causes, that experience and experiment are the basis of nature study, and that mathematics is the key to knowledge of nature.
   6. Nevertheless, while some of the founders of modern science were academics, most were not, nor was the university a center of modern scientific research until the 19th century.
   7. Galileo and Newton, for example, were university professors, but Francis Bacon, René Descartes, Robert Boyle, Robert Hooke, Christiaan Huygens, and Gottfried Leibniz were not.

Essential Reading:

Edward Grant, *Physical Science in the Middle Ages*.

Charles Homer Haskins, *The Rise of Universities*.

Questions to Consider:

1. Why didn't the Catholic Church suppress the manifestly secular new universities?
2. Why didn't modern science emerge in the 14th century?

Lecture Eight
The Medieval Revolution in Know-How

Scope: The invention of the university was symptomatic of a new dynamism that characterized western Europe in the period 1100–1350: Gothic cathedrals, expanded trade extending to India and China, urbanization, and technological innovations of enduring influence. Innovations dramatically raised agricultural productivity; improved ship design; and via new nautical charts and the use of the compass, stimulated expanded trade. Water- and wind-powered mills became ubiquitous, followed by a wide range of water- and wind-powered machinery. The need for complex gear trains and the means of transferring power led to the invention of the weight-driven clock, the cam, the clutch, mechanical automata, and the dissemination of mechanical skills in the population. Expanded trade led to new financial techniques and institutions, among them, banks and corporations. In effect, western Europe experienced an industrial revolution as well as a cultural renascence, revealing the role played by society in determining the influence on it of innovations.

Outline

I. The 12th-century renascence was unquestionably a time of intellectual rebirth, but philosophical ideas were only symptomatic of a much deeper cause with wide consequences.

   A. It can be argued that Europe experienced its first "industrial revolution" in the 12th century, not the 19th century.
   1. The invention of the university was one symptom of a rebirth of social and cultural dynamism in the Christian West.
   2. Another symptom was a sudden, dramatic increase in the use of machines and in technological innovation generally.
   3. One can legitimately call this an industrial revolution, but it is not an expression of the Graeco-Roman idea of techno-science.
   4. This industrial revolution, like the first phase of the more famous 19th-century one, was closer to the tinkering/inventiveness of know-how in prehistory.

B. The use of water- and of wind-powered mills is a dramatic illustration of this medieval phenomenon.
1. Roman "engineers" were quite familiar with water-powered mills for grinding grain, but only a few such mills were built before the empire was overrun.
2. By contrast, when William the Conqueror ordered a survey of English assets shortly after his victory at Hastings in 1066, his surveyors counted thousands of water mills in use.
3. This degree of exploitation was matched elsewhere in western Europe and, shortly thereafter, in northern Europe and intensified in the period 1100 to 1400.
4. Although initially used for grinding grain, mills were soon adapted to powering machines for making paper out of rags, for operating the bellows of iron forges, for cutting wood and stone, and for hammering, fulling, and tanning.
5. Building and operating these mills required significant capital investment for the period and led to the formation of the first shareholder-owned corporations, some of which still exist after more than 700 years.
6. Starting probably in the 12th century and certainly in the 13th, windmills began to be exploited intensively, both for grinding grain and for pumping water.
7. The British were noted for developing windmill technology, but the Dutch became its masters for pumping water on a large scale to drain swamps and low-lying land.

C. One value underlying this embrace of industrial technology is secularism.
1. We saw this as a value underlying the university movement in the assertion of the autonomy of reason vis-à-vis faith and the attribution of value to the pursuit of secular philosophy and knowledge of nature.
2. Here, secularism finds expression in the value attributed to working hard, to wealth, and to labor-saving devices.

D. The volume of machine utilization had broad and deep implications for medieval society.
1. The construction of so many mills implies an extremely broad base of mechanical skill.
2. The centuries-long stream of innovations in mill designs and in the construction, efficiency, and application of mills is

expressive of the depth of the medieval engagement with technology, implying continual experimentation in the search for a technological "edge."
3. All these applications required mastering gear-train design and construction, as well as power takeoff, transmission, and control mechanisms, for example, the clutch and the cam, lantern and crown gears, and gearing for speed and power.
4. Furthermore, a mill construction industry of this magnitude creates a massive demand for hardware on a scale that forces innovation in metal cutting and fabrication and a demand for standardization.

II. By far, the most famous popular invention of the period in question was the weight-driven mechanical clock.
   A. Mechanical clocks, in the form of water clocks, were common in the Graeco-Roman and Islamic periods.
      1. These clocks evolved into an application of science to technology as Archimedes had developed a mathematical theory for the physics underlying their design.
      2. The weight-driven mechanical clock, which for centuries was no more accurate than well-made water clocks, triggered a totally different social response in late-medieval Europe from the water clock.
      3. Technologically, this new clock is indebted to the body of gear-train knowledge that grew up around watermill and windmill machinery.
      4. The key innovation is the verge-and-foliot escapement that checks the falling weight and uses its downward motion to drive the time display (for centuries, just an hour hand) and often complex automata.
      5. The higher the clock mechanism was mounted, the longer the fall; thus, these clocks became public time displays located in monumental structures, such as campanile.
      6. With the clock came the standardization and commercialization of time, but this surely reflects existing social values, not the causal consequences of the clock as an artifact.
   B. As a matter of fact, the clock has been perceived by many historians as the "defining technology" of the late Middle Ages.

1. The clock, it is claimed, secularized time and, in the process, secularized the temporal organization of social life, while previously, social life was organized around the religious "hours" of the Catholic Church.
2. The point at issue is technological determinism, that is, whether a new technology causes a society to adapt to it, if necessary at the expense of prevailing values, or whether prevailing values determine the way a society responds to a new technology.

C. Neither the clock nor watermills and windmills, however, were isolated instances of Western society opening technological innovation "doors," then rushing through them.
   1. Innovations in agricultural technology and a suddenly favorable climate raised farm productivity even as the population increased dramatically.
   2. These innovations included a "heavy," iron-tipped ploughshare capable of working the denser soils of central and northern Europe, replacing the "light" plow inherited from the Romans.
   3. The introduction of the breast harness allowed the use of horses as draft animals, replacing the much slower oxen.
   4. The more widespread adoption of fertilization and a three-field crop rotation system, "scientific" farming for the time, also increased yields.
   5. The higher agricultural productivity seems related to the rapid growth in urban populations, reinforcing commerce and the university movement.

D. Concurrently, organized long-distance trade increased dramatically, aided by marine technology innovations.
   1. A centrally mounted stern-post rudder, with a complementary steering system, that allowed ships to sail in deeper waters far from shore was in use by the 13th century, and soon after, so-called lateen sails replaced square rigging, dramatically improving maneuverability.
   2. The magnetic compass was in use by the 12th century, and more sophisticated seaport charts increased the reliability of navigation.
   3. Contact with Moslem merchants in North Africa and the Middle East increased the scope of trade and stimulated the

adoption of managerial technologies that reinforced expansion of trade, for example, Hindu numerals, double-entry bookkeeping, and letters of credit.
4. One clash of values that arose in this context was the religious prohibition against usury and the growing need for capital by merchants and entrepreneurs.
5. Ultimately, the social pressure for economic growth forced a redefinition of usury to allow making money from money!

E. These developments illustrate what Lynn White meant by saying that innovations only opened doors for a society.
1. This view of White's, which puts responsibility for responding to an innovation on society and its values, contrasts with technological determinism.
2. On this latter view, new technologies impose themselves on society in ways that derive from characteristics of the technologies.
3. The Portuguese used new nautical technologies to sail to Asia and seize the spice trade.
4. This inaugurated a centuries-long period of European colonization and exploitation that literally transformed the world.
5. Western and southern European societies rushed through the doors opened by the available and newly created technologies of the 12th and 13th centuries, creating a social environment favorable to secular knowledge, industry, and commerce.

Essential Reading:

Jean Gimpel, *The Medieval Machine*.

Lynn White, *Medieval Technology and Social Change*.

Questions to Consider:

1. Why, after centuries retrospectively called the Dark Ages, did European culture suddenly become dynamic in the 12th century, and why only western Europe?
2. How did the Crusades reflect and affect this dynamism?

Lecture Nine
Progress Enters into History

Scope: The aggressive embrace of the idea of progress by Western societies was a symptom, not only of the openness of those societies to be transformed, but of their *active pursuit* of social transformation. Explicitly, the idea of progress was a product of the early Renaissance, introduced by poets, artists, and scholars of the Humanist movement, who invented the related idea of a rebirth of cultural excellence. Implicitly, however, the 12th-century industrial revolution and cultural renascence are evidence that the active pursuit of social transformation by western European societies was already in place. Furthermore, the social transformation being pursued was secular, in spite of the contemporary power of the Catholic Church. Gothic cathedral construction notwithstanding, people were pursuing wealth, comfort, pleasure, personal freedom, and civil power. The idea of progress is, thus, a secular idea that paved the way for secular modern science to become an agent of social reform based on scientific reason.

Outline

I. The idea of progress is not a scientific idea, but it is an concept that became identified with the idea of science and with modernity at a certain point in European history.

   A. If history is logical, the logic certainly is elusive, as reflected in the fact that modern science did not emerge in the 14th century.
   1. There have been theories of history, notably that of G. W. F. Hegel, in which events unfold deterministically according to universal laws.
   2. Aristotle had argued that history was not logical in the deductive sense because it was particular and, thus, could not be a science.
   3. All the ingredients for the rise of modern science seemed to be at hand in the 14th century, including the value placed on secular knowledge, but the cultural dynamism of the preceding centuries waned.

4. Attributing this waning to the black plague sounds plausible, but this explanation is not convincing, in part because the idea of progress first emerges in the 14th century!

B. The idea of progress first gained currency in Europe within the Humanist movement of the early Renaissance.
1. The 14th-century Italian poet Petrarch embedded the possibility of progress into the Humanist movement that he instigated, mentored, and inspired.
2. Petrarch proclaimed the prospect of a "renaissance" of the cultural excellence achieved by the Classical-era Greeks and Romans.
3. Petrarch periodized European cultural history into a glorious Classical Age, a Dark Age, and the dawn of a rebirth of the Classical Age.
4. The rebirth was signaled by the emergence of new Italian poetry, especially Dante's *Divine Comedy* and Petrarch's own work.

C. Petrarch was concerned primarily with language and literature, not with politics, social institutions, philosophical knowledge, or technology.
1. Petrarch's life extended over the first outbreak of plague in 1347–1348; thus, physical well-being was not central to his idea of progress.
2. Nor did he see the university as we do or think that windmills and clocks heralded a new age.
3. Petrarch was extremely sensitive to language, and he abhorred the debasement of spoken and written Latin over the previous 1000 years.
4. But now, he thought, the times were changing, and people were emerging who were committed to pursuing literary excellence.

D. Petrarch argued that the path to excellence began with imitation of the greatest literature ever written, namely, that of the Classical Age.
1. Such imitation required recovery of great literary texts and their close study so that their style could be emulated.
2. Emulation, however, was only a means to the end of rising above the literary achievements of the past, and here, the idea of progress makes its debut.

3. Progress is possible if we assimilate the best of the past and build on it creatively: The goal is to be better than the past.

II. The Humanist movement has often been dismissed as a pretentious nostalgia trip that impeded modernity, but this is a serious error.
   A. Emulation, recall, began with the recovery and close study of the "best" literary works, those of Classical antiquity.
      1. By 1300, hundreds of texts had been translated into Latin, but overwhelmingly, these translations were of very poor quality.
      2. Typically, translations that fed into the university "market" were of Arabic books, which were themselves translations of translations most of the time.
      3. The ideal, of course, would be to have the original Greek or Latin text to study closely, but the originals were all gone.
   B. The Humanists, inspired by Petrarch, began a centuries-long search for ancient manuscripts.
      1. The targets were monastic libraries, private libraries, Moslem scholars and merchants, and Byzantium.
      2. They succeeded in finding thousands of manuscripts that were copies of Greek- and Roman-era originals, including many in natural philosophy, medicine, mathematics, and technology.
      3. Although the Humanists were often not interested in the content of technical works for their own use, they were committed to publishing accurate original editions of them.
   C. The generic problem was that the copies the Humanists found were invariably corrupt.
      1. The Humanists responded by developing techniques for recovering accurate original texts from corrupted copies.
      2. One family of techniques involved manuscript collation if there were multiple copies of the same manuscript, as was often the case.
      3. A second technique was manuscript emendation, which was much more challenging than collation.
      4. Emendation required understanding both the content of a text and the original language so well that corrupted words could be recognized and replaced and lacunae filled in.
      5. This project revealed the historicity of language—that words and scripts changed their meanings over time—and the need for specialist dictionaries.

D. The search for manuscripts began in the late 14th century, intensified in the 15th, and extended through the 16th century.
   1. Fortuitously, the Humanist project of recovering accurate texts of great works of literature, philosophy, mathematics, and "science" overlapped the introduction of printing.
   2. Printing dramatically improved the ability of the Humanists to publish and distribute recovered texts; it also held the promise of underwriting the costs of scholarship or enabling a life as an independent scholar.
   3. Most important, it made more effective the Humanist organization of a network of interacting, sometimes collaborating scholars, reviewing and critiquing one another's work, constructively and destructively, in progress or after publication.

E. In the process of pursuing their own interests, the Humanists made the idea of progress an issue for Renaissance Europe, and they provided valuable intellectual "tools" for modern science.
   1. They published, for the first time, accurate printed editions of scores of texts by Classical mathematicians, "scientists," philosophers, and technologists, correcting countless errors and making previously unknown texts available.
   2. Vitruvius's *On Architecture*; Ptolemy's *Geography*; the complete works of Plato; several treatises by Archimedes, including *The Sand Reckoner*; Apollonius of Perga's *The Conic Sections*; and works by Hero of Alexandria were profound influences on modern science.
   3. The Humanist challenge of using critical methodological "tools" to recover an original text from corrupted copies is strikingly similar to the challenge of modern science: using the "scientific method" to discover the reality behind "corrupt" sense experience.
   4. The Humanist creation of informal networks of scholars and formal associations, sharing ideas and work in progress and exposing their finished work to peer criticism and approval, was reflected in the way modern science was practiced, including the formation of scientific societies.

III. The idea of progress is so central to modern life that we take it for granted as a fact or a self-evident truth.

A. The case for progress changed dramatically in the 15th and 16th centuries.
   1. Petrarch's idea of progress was, as we have seen, narrowly conceived, but it put the idea in people's minds.
   2. In the course of the 15th century, the idea of progress was explicitly coupled to new knowledge, especially mathematical and "scientific" knowledge, and to new technologies.
   3. The dynamism of the 12th and 13th centuries again animated European societies, and this time, the new technologies were based on knowledge, primarily mathematical knowledge.

B. The case for progress was much broader and much stronger by the end of the 16th century, but progress was still a controversial claim to make.
   1. A common theme in the Renaissance was the "battle" of the ancients and the moderns, a battle fought in words mostly but often in various forms of spectacle/entertainment.
   2. On the side of the moderns were the compass, the new printing technology, paper, maps, and voyages of global discovery.
   3. The idea that people were better—wiser—in the 16th century than in the distant past was rejected by many who held that the case for progress was superficial.
   4. By the 17th century, the "battle" had been won by progress, which became associated with reason and, thus, with science.
   5. The proclamation of the Age of Reason in the 18th century was based on a belief that science could be an agent of reform in human affairs.
   6. The maturation in the 19th century of the idea of techno-science certainly *changed* the human condition, but has technological progress *improved* that condition?

Essential Reading:

Donald R. Kelley, *Renaissance Humanism*.

Robert Nisbet, *A History of the Idea of Progress*.

Questions to Consider:

1. What accounts for the emergence and embrace of a secular idea of progress in societies dominated for centuries by spiritual ideas and values?
2. Have science and technology delivered on the promise of the idea of progress that the products of reason would improve the human condition?

Lecture Ten
The Printed Book—Gutenberg to Galileo

Scope: Printed texts were not new in the West in the 15th century and were more than 1000 years old in China. Printing using movable type was also an old technology when Gutenberg introduced it to Europe, but the response of European societies to his print technology perfectly illustrates the thesis that innovations only open doors; social factors determine if a society enters through them. The impact of printed texts on Western societies was dramatically different from its impact in China or Islam. European culture became "print drunk" instantly. By 1500, some 10 million texts, including 2 million books, had been printed at a time when the literacy rate was very low. The response triggered the creation of a vast system to supply, produce, and distribute texts, and new institutions were created to protect and reward producers and increase literacy, promoting further increases in text production and distribution.

Outline

I. The printed book is an excellent illustration of how technological innovations change the way we live.
   A. Books are only the most visible manifestation of a largely invisible, highly complex social-technological system at work.
      1. This is true of artifacts generally, not just of books.
      2. The visibility of artifacts distracts attention from the underlying system of which they are manifestations.
      3. Although artifacts are described as causes of social change, social factors play a central role in deciding if artifacts are to be produced, how and in what form, and how they are to be distributed and used.
   B. If an invention such as Gutenberg's printing press only "opens a door," we need to ask what it takes for a society to "cross the threshold."
      1. Gutenberg did not invent printing.
      2. The Chinese were producing block-printed texts by the 6th century, eventually producing multi-block, multi-color texts.

3. In the 11th century, a Chinese printer developed movable (fired) clay type, and in the 14th century, printers began using movable wooden type, early experiments with metal type having failed.
4. Between 100 B.C.E. and 105 C.E., paper was invented by the Chinese using vegetable matter as the "pulp," and papermaking technology spread to Korea, Japan, and the Middle East.
5. These Chinese print technology innovations spread via Korea to Japan by the 12th century and, by 1400, to the West.
6. The first European paper mills, for example, were built in Islamic Spain in the 12th century and quickly moved north into France, where the pulping process was water-powered, the paper now being rag-based.

C. The invention that triggered the 15th-century book "revolution" was high-volume printing using movable metal type.
1. Gutenberg was the first successful integrator of movable metal (lead) type, a modified ink with an oil base, and an adaptation to printing needs of oil and linen presses.
2. Oil-based paints for artists were introduced at about this time in Europe by Jan van Eyck.
3. The impact of this new print-production technology, given the population of Europe and the low literacy rate, was astonishing: By 1500, 2 million books alone had been printed, altogether some 40,000 different titles!
4. Echoing Lynn White's dictum, western European society burst through the door opened by this new technology, but what was required for it to do so?

II. The major elements of the iceberg/system of which the printed book is the "tip" are easily identified.

A. Exploding in just 50 years from a low-volume specialty item to a mass-production commodity, book production required a comparable increase in the production of paper and ink.
1. Scaling production up to the new level demanded by the printing press, however, required a whole new kind of paper industry: more and larger mills, more raw materials, more workers, more investment, increased transportation capacity,

scaled-up marketing and distribution, and more complex business organization.
   2. All these factors apply to supplying appropriate inks, as well.
   3. The print shop needed to be "fed" as the number of presses grew, as markets grew, and as competition increased.
   4. Skilled workers were needed to build, operate, and maintain presses and auxiliary equipment.
   5. Typecasting was a new job category, as was typeface design, and the demand for lead rippled back to increased mining.
   6. Even when printing existing texts, printers needed educated people to prepare the content and copyeditors to check for errors and monitor the accuracy of the production process.
   7. Very soon, printers needed authors to provide original content.
   B. The printing press also needed to be "fed" and protected at the demand end of the book-production process.
      1. A book-*selling* industry—more broadly, a printed-materials-selling industry—was needed that matched demand to the scale of the printed-materials-*production* industry.
      2. Printers needed distribution channels that could move inventory more rapidly than medieval-era trade fairs.
      3. As the market grew, bookselling became a business in its own right, with an interest in playing printers off against one another.
      4. From the beginning, printers, authors, sellers, and the reading public needed protection from "pirated" texts, especially editions that were not accurate replicas of the author's original text.
      5. New markets were needed to absorb growing production capacity, which required dramatic increases in literacy.
      6. A "literacy industry" arose to increase the scale of education and make more people want to read more.
      7. Growth in demand fed back into the production system, promoting further growth!
III. Everyone acknowledges that modern printing exerted a profound impact on Western societies, but why it had that impact and how that impact was effected are little understood, as is the impact of modern printing on modern science.

A. The book as the visible "tip" of a largely invisible socio-technic system reveals an important feature of all such systems.
  1. There was no central planning for the creation of this system.
  2. No one person or institution orchestrated the complex supply, production, distribution, and demand channels.
  3. There is a lesson here in self-organization that we will return to in Lecture Thirty-Four.

B. To understand the impact of printing on Western society, we need to look at the demand side and the content of printed works.
  1. The question of *what* to print was initially answered by printing Bibles, religious literature, and Greek and Latin classics.
  2. The Humanists became a source of new editions of these classics, but as demand grew, so did the need for original material.
  3. As the investment in mass production grew, so did the need for mass "consumption."
  4. The size of the market is directly related to the literacy/education rate of the population.
  5. Francis Bacon, "father" of the experimental method, was primarily an education reformer, and universal literacy was one of his objectives.

C. The response after 1450 to the possibility of mass production of printed material is revealing in other ways, as well.
  1. A positive feedback loop, reminiscent of the impact of the invention of writing, was created, so that production and consumption became mutually reinforcing.
  2. The invention of writing was immediately followed by the appearance of vast numbers of documents, as if a dam had burst. So it was with printing.

D. The power and attraction of mass production of texts are not as obvious as they may seem.
  1. One view has been that printing authoritatively "fixes" a text, so that authors can be assured that every reader is reading exactly the words they wrote and looking at exactly the same illustrations they provided.

2. In fact, others have argued, this was not at all the case, because of the new ease and commercial attractiveness of plagiarism and pirated versions of "bestsellers."
3. Galileo, for example, carefully produced a "deluxe" edition of a slim book that he hoped would make his fortune, *The Starry Messenger*, describing his telescopic discoveries. Within a year, pirated copies with distorted illustrations were being sold all over Europe.
4. The great Danish astronomer Tycho Brahe attempted total control over the printing and distribution of his books by owning or controlling every step in the process—in spite of which he failed.
5. The new technology thus required the creation of new social institutions to deal with these problems and with the problem of controlling morally, politically, and religiously "subversive" texts.

Essential Reading:

Elizabeth L. Eisenstein, *The Printing Revolution in Early Modern Europe*.
Adrian Johns, *The Nature of the Book*.

Questions to Consider:

1. How does a social-technological system, such as the print industry, come into being without central planning and coordination?
2. What was it about 15th-century Western culture that caused a demand for universal literacy in response to the new print technology?

Lecture Eleven
Renaissance Painting and Techno-Science

Scope: Mathematical physics was the heart of modern science when it was created in the 17th century, and it remains so. That mathematics is the "language" of the reality behind experience was a Pythagorean teaching in antiquity, but it acquired cultural force in the West during the Renaissance in advance of modern science and at the hands of artists, engineers, and innovators in mapmaking, navigation, and commerce. The rediscovery of central-vanishing-point perspective drawing, for example, was immediately and universally acclaimed as the means to "realistic," truthful representation. The application of abstract mathematics to a wide range of practical problems reinforced the correlation, if not the identification, of mathematics with reality. The origins of modern engineering lie in these applications, again, in advance of modern science. They also stimulated the study of mathematics and the production of original mathematics in Christian Europe for the first time since the decline of the Roman Empire.

Outline

I. The Humanists, promoting their own agenda, introduced the idea of progress into European society, which then linked progress to technological innovation and scientific knowledge; Renaissance artists contributed significantly to the resurrection of the idea of techno-science!

   A. The Florentine architect-engineer Filippo Brunelleschi's demonstration circa 1415 of central-vanishing-point (CVP) perspective drawing sparked an almost immediate transformation of European art.
   1. For 1000 years, the pictorial art of Catholic Europe did not exploit perspective drawing.
   2. The technique is referred to in general terms (*skenografia*) in Vitruvius's *On Architecture* as if it were familiar to practitioners, and the frescoes unearthed at Pompeii reveal that it was indeed familiar 100 years later.

3. Centuries before that, Plato had written critically of the painter Zeuxis, much praised for painting so realistically that birds attempted to eat the grapes he depicted.
4. Nevertheless, paintings in the Byzantine, Romanesque, and Gothic periods were overtly two-dimensional.

B. It is likely that CVP perspective drawing was deliberately rejected by early medieval artists and later became unfamiliar through disuse rather than from ignorance or incompetence.
1. Perspective drawing is valuable because it allows the artist to depict things as they appear to the physical eye, as they are "naturally." If the goal of the artist is to represent people, objects, and scenes as we experience them visually, "realistically," then CVP perspective drawing is essential to painting.
2. If, however, the goal of the artist is to use the content of a painting to arouse in the viewer an ideal intellectual or religious response, in the belief that the ideal is superior to, because more real than, the natural, then CVP perspective drawing is not only *not* valuable, but it is a corruption.
3. Plato thus criticized the painter Zeuxis for using his talent to fool the eye, which led people to identify the real with the material, and this was certainly the sentiment of medieval Christian painters.
4. Recall that throughout the Middle or Dark Ages, there was virtually no secular art in Catholic Europe.
5. It follows that medieval artists put no value on naturalism or on individuality/particularity, which is a mark of the material, and thus, they could put no value on CVP perspective drawing even if they knew of it.

C. Brunelleschi seems to have rediscovered CVP perspective on a visit to Rome and demonstrated it on his return to Florence by displaying an experimental painting he made of the baptistry in Florence.
1. Around 1425, Masaccio employed the technique in a masterful series of frescoes that attracted great attention.
2. Shortly thereafter, the painter Piero della Francesca, who was also a mathematician, wrote the first of what would be many handbooks for artists explaining the mathematical technique

for foreshortening that underlies CVP perspective and makes "realistic" drawing possible.
3. The triumph of this new technique was so complete that we identify Renaissance painting with its use.
D. To embrace perspective drawing as the only way to "truth" in art is to redefine reality for art.
1. CVP perspective drawing allows the artist to depict three-dimensional objects on a two-dimensional surface such that the eye (the mind, of course) responds to the flat visual representation as it does to the material visual object.
2. We should be astonished by the cultural phenomenon that Brunelleschi's introduction of this technique inaugurated: Within 50 or so years of his experimental painting of the baptistry in Florence, CVP perspective drawing was ubiquitous, adopted by artists and public alike as the *only* way to paint.
3. To value CVP perspective so highly is to value the natural as real, to embrace the corporeal as real, to identify truthfulness with accurate representation of the natural, the material, *and the individual*, overturning 1000 years of aesthetic, philosophical, and religious values.
4. Fifteenth-century Europe remained thoroughly Catholic at one level, but the secularism, naturalism, and individualism that had surfaced in the 12th-century renascence now were revealed to have a much deeper hold in society.
5. One manifestation of this hold was the commissioning even by prominent Church officials of paintings depicting scenes from Greek mythology.
6. Another was the aggressive egotism of Renaissance master artists, documented by Giorgio Vasari.

II. CVP perspective drawing was immediately understood to be a technique of mathematics, which was thereby coupled to 15th-century art and its values.
A. That the power of CVP perspective drawing to "capture" nature was a consequence of following a mathematical formula implied that mathematics was intimately connected to nature.

1. This connection echoes the Pythagorean metaphysics that made mathematical form the essence of material objects and the order of nature.
2. Making mathematics the basis of natural philosophy (knowledge of nature) was part of the curriculum at leading medieval and Renaissance universities.
3. Ptolemy's mathematical astronomical theory was required reading for virtually all students, but the recovery by Humanists of his *Geography* exerted a new influence.
4. The *Geography*, whose original maps were lost, described the world as Ptolemy knew it but also the use of mathematics to represent the three-dimensional surface of the Earth accurately on a two-dimensional surface.

B. Throughout the 16th century, the idea that mathematics captured essential features of the real material-physical world spread in concrete ways.
1. Ptolemy's *Geography*, coming at a time when Europeans were embarking on global voyages of exploration, stimulated the rise of schools of mathematical cartography, one exponent of which was Gerard Mercator, famous for his Mercator projection maps.
2. John Dee was an influential 16th-century British mathematician who taught techniques of mathematical navigation to many British ship pilots sailing to the Americas and Asia.
3. From the 15th century on, rival mathematical theories of musical harmony, based on or deviating from Pythagoras's original insight into tone systems, played a central role in music composition and musical instrument design and performance.
4. Galileo's father, Vincenzo Galilei, wrote two books on this subject, rich with experiments of his own design, and Kepler's greatest work in his own eyes was his 1619 *Harmony of the World*, coupling music and astronomy via mathematics.

C. In the course of the 16th century, practical applications of mathematics reinforced the connection to nature forged in painting and created a pillar of the 17th-century Scientific Revolution.

1. By the end of the 16th century, books of machine designs were an established publication genre, and machine designers adopted and adapted CVP perspective drawing.
2. This had the effect of making the depictions of machines more "realistic" and visually intelligible, but they were also made more intelligible mechanically by including cutaway and exploded drawings of details of the mechanisms.
3. This new capability seems to have stimulated the design of more complex machines, because these could now be depicted convincingly and in a mechanically intelligible way.
4. The engineering version of the mathematics underlying CVP perspective drawings is *orthographic projection*—a two-dimensional representation that allows reconstruction of a complex three-dimensional object from that representation.
5. This capability developed in 16th-century machine illustration and achieved formal expression—another linkage of know-how to knowledge—in projective geometry, co-invented by Girard Desargues and Blaise Pascal circa 1640.
6. A fascinating illustration of the new capabilities of Renaissance engineering is Domenico Fontana's account of his contract to move an ancient Egyptian obelisk weighing more than 700,000 pounds some 260 yards to make way for the construction of St. Peter's Cathedral in the new Vatican.

Essential Reading:

Samuel Y. Edgerton, Jr., *The Heritage of Giotto's Geometry*.

Bertrand Gille, *Engineers of the Renaissance*.

William Barclay Parsons, *Engineers and Engineering in the Renaissance*.

Charles Singleton, *Art, Science, and History in the Renaissance*.

Questions to Consider:

1. What do we mean when we call a calculatedly deceptive simulation of the appearance of the subject of a painting "realistic" and "truthful"?
2. Did the monotheistic religious tradition of the West contribute to the scientific idea of an abstract reality behind concrete experience?

Lecture Twelve
Copernicus Moves the Earth

Scope: The idea that the Earth is motionless at the center of a large but finite Universe that revolves around the Earth matches our commonsense experience of the motions of the Sun, Moon, planets, and "fixed" stars. By contrast, we have no sensation whatsoever of any motion of the Earth. Furthermore, that the Earth is the center of the Universe fits very well with religious and mystical philosophical teachings that the Universe was created for the sake of mankind and that human beings play a central role in a cosmic drama of creation and salvation/damnation. Ptolemy's mathematical model of such an Earth-centered Universe thus anchored a conception of the Universe as both manifestly and anthropocentrically ordered. Copernicus's theory, in addition to violating common sense by attributing multiple concurrent motions to the Earth, undermined the anthropocentricity of the cosmic order and, ultimately, the manifest orderliness of the cosmos itself.

Outline

I. Copernicus did not just disturb the Renaissance Universe, but he redefined it, and after 150 years, his new definition stuck but only for 200 years.

   A. The Renaissance Universe incorporated two overlapping but distinctive lines of thought, one astronomical, the other religio-metaphysical.

      1. The most vivid expression of the latter was Dante's epic poem subsequently named *The Divine Comedy*.

      2. Literally, the poem recounts a descent into hell within the Earth, followed by an ascent from the Earth through the physical heavens to the abode of the saints in God's presence.

      3. Throughout, Dante displays a sophisticated grasp of contemporary astronomical theory, which he used selectively for poetic effect, but what is most important about the astronomy in the poem is the symbolism.

4. Symbolically, the surface of the Earth, on which the human drama is played out, is poised midway between hell and heaven.
5. The descent into hell is the mirror image of the ascent to heaven, which emphasizes the idea that the ultimate order of the cosmos is centered on mankind and his spiritual destiny.

B. Parallel to Dante's religious-symbolic expression of the centrality of the Earth in the Universe, there was the dominant "professional" Earth-centered astronomical theory: Ptolemy's.
1. Ptolemy's theory reigned for some 1,400 years, but he never intended it to be understood as physically real or physically true.
2. His theory was strictly mathematical, its purpose to deduce the observed but merely apparent motions of the Sun, Moon, and planets from their real motions.
3. These "real" motions had to satisfy Pythagoras's criteria, which on mathematical-metaphysical grounds, required that heavenly bodies move only in circular orbits at uniform speeds.
4. Ptolemy's model is extremely complicated, but it works quite well for naked-eye observations.
5. That is, using his theory, one can calculate in advance where each planet will be in the night sky as a function of time with an accuracy of about 1/6 of a degree.
6. Note well: The order of the Universe for theoretical astronomers is mathematical, hence, accessible only to the trained mind, visible only to the mind's "eye."
7. The Universe visible to the body's eye is only apparently orderly but in fact is confused, misleading, and disorderly.

II. This was the Universe that Nicolaus Copernicus inherited, then transformed.

A. Copernicus was theologically orthodox but intellectually radical.
1. Copernicus was a university-trained Polish Catholic priest who then spent years studying in Humanist Renaissance Italy before returning to Poland and a career in the Church.
2. Reading Archimedes, he found a reference to an idea of Aristarchus of Samos that the Sun was the center of the

cosmos and the Earth just another planet (an idea treated dismissively by Archimedes!).
3. Some ancient, some Islamic, and some late-medieval Christian writers referred to this or similar ideas, but no one had acted on them: Copernicus did.
4. Copernicus devoted himself to astronomy and, by 1512, was committed to working out a theory in which the Sun was stationary at the (near-)center of the Universe, while the Earth moved around it on a circular path at a uniform speed, as Pythagoras's mathematical metaphysics required.

B. The practical motivation for such a theory, quite apart from the philosophical motivation, was calendrical.
1. Calculating in advance the positions of the planets in the night sky served important "industries": almanacs for farmers, travelers, and sailors; and charts for both astrologers and astronomers.
2. A byproduct of these calculations was more accurate determination of the length of the solar year, which was of great importance to the Catholic Church in setting the date of Easter.
3. Copernicus enjoyed a growing reputation as an astronomer when the Church was forced to recognize that Julius Caesar's calendar reform of 46 B.C.E. caused Easter to come far too early.
4. A papal commission was appointed that led to the Gregorian calendar reform implemented in Catholic countries beginning in 1582, later in Protestant countries, and still not completely by Russian Orthodox communities even today.

C. In 1543, the year he died, Copernicus finally published his theory in his great work *On the Revolutions of the Heavenly Spheres*.
1. He was goaded into publishing by a young astronomer named Rheticus, who learned of Copernicus's radical new theory and came to him to learn it.
2. This book works out in full mathematical detail an alternative to Ptolemy's theory, one in which the Earth moves around the Sun.

D. Although Copernicus is universally hailed as the founder of modern astronomy and, in a sense, he was, his theory of the

Universe and even of (what we call) the Solar System is completely wrong.
1. First of all, he had virtually no observational data that had not been available to Ptolemy or that could not have fit perfectly into Ptolemy's theory.
2. Thus, Copernicus's theory is an alternative interpretation of the same data available to Ptolemy's theory: The data do not distinguish between the two theories.
3. Second, the planets do not move in circular orbits or at uniform speeds.
4. Third, there is no pre-telescopic observational evidence to support the claim that the Earth moves around its own axis in 24 hours or around the Sun in a year. On the contrary, observational evidence works against the theory.
5. Finally, because he accepted circular orbits and uniform speeds and had to place the Sun slightly off-center, Copernicus, like Ptolemy, had to use epicycles. For this reason, his theory is almost as complicated to use for calendrical calculations as Ptolemy's and no more accurate: both are off by about 10 minutes of arc.

III. Copernicus, almost certainly unintentionally, instigated a cultural revolution at the metaphysical level, as well as an astronomical revolution.
  A. Unlike Ptolemy, Copernicus insisted that his theory was physically real.
  1. Copernicus's book appeared with a preface describing his theory as mathematical only. Ostensibly, this preface was by Copernicus, but in fact, it was written by a Protestant editor seeking to protect Copernicus from a charge of heresy.
  2. The assertion that the theory was mathematical only would allow for the Earth "really" to be stationary, the position the Church insisted was mandated by the Bible.
  3. But Copernicus's text does not support that reading, and subsequent Copernicans, such as Kepler and Galileo, took the theory to be intended as a mathematical expression of physical truth.

B. In spite of being wrong from the perspective of the subsequent history of science, Copernicus's theory occasioned a re-conception of the Universe and the Earth's place in the Universe.
   1. Again using Lynn White's phrase, Copernicus opened a door for Western intellectuals, who were so anxious to get through it that they modified the theory freely.
   2. Where Copernicus thought that the Universe was compact, prominent Copernicans proclaimed it to be infinite.
   3. The symbolic significance of removing the Earth from the center of the Universe is profound but becomes still more profound if the Universe is infinite.
   4. Many professional astronomers used and taught Copernicus's theory, thus disseminating his ideas, but withheld support for its physical truth.
   5. Kepler abandoned circular orbits and uniform speeds, which made the orderliness of the Universe still more abstract and intellectual than in Pythagoras's original version.
C. Kepler and Galileo were the leading champions of Copernicus's theory, which remained a minority view until Newton's *Principia Mathematica*.
   1. Kepler fundamentally altered Copernicus's theory, introducing forces that bound the planets to the Sun and caused them to rotate as they did, but he still considered himself to be a follower of Copernicus.
   2. Tycho Brahe rejected Copernicus's moving Earth and developed a theory of his own that was preferred by many if not most educated people well into the 17th century.
   3. Kepler worked as Brahe's assistant, ostensibly to work out the mathematics of Brahe's theory but, in reality, to get access to Brahe's new observational data.
   4. Using instruments of his own design, Brahe compiled the most accurate planetary position observations ever made. When Brahe died, Kepler seized the data and used them to support his own version of Copernicus's theory.
   5. Galileo was an "orthodox" Copernican, rejecting Kepler's modifications (which turned out to be correct!) and Brahe's theory, even though it fit his telescopic observations as well as Copernicus's.

Essential Reading:

Nicolaus Copernicus, *On the Revolution of the Heavenly Spheres*, book 1.

Thomas Kuhn, *The Copernican Revolution: Planetary Astronomy in the History of Western Thought.*

Questions to Consider:

1. What are the implications for scientific theories of the fact that Copernicus based his theory of the heavens on essentially the same facts Ptolemy used?
2. In the absence of any empirical evidence whatsoever for the Earth's motion, either on its axis or around the Sun, why did anyone adopt Copernicus's theory?

Lecture Thirteen
The Birth of Modern Science

Scope: One enduring mystery of modern science is why it happened where and when it did, in the Christian culture of 17th-century western Europe. Had modern science emerged in late Graeco-Roman antiquity, in 11th- or 12th-century Islam, or in China anytime after the Tang dynasty, there would be no mystery. One clearly crucial development was the idea that knowledge of nature can be developed only by applying a radically impersonal method to personal experience, a method that generates accounts of experience to be validated by observation, prediction, and control and replication by others. In truth, no one method was common to all the founders of modern science. Francis Bacon claimed that the only correct method was based on induction; René Descartes, that it was based on deduction; and Galileo Galilei, that it was based on a fusion of experimentation and deduction, following the model of Archimedes. All agreed, however, that method was central.

Outline

I. Modern science aspired to being a natural philosophy, which implied having knowledge of nature, not just opinions or beliefs.
   A. The Scientific Revolution was more evolutionary than revolutionary.
      1. Many of the ingredients that went into the making of modern science were at hand at the turn of the 17th century.
      2. These were supplemented by contributions from Islamic culture, the medieval university, and the 12th-century "renascence."
      3. The 15th and 16th centuries were extraordinarily dynamic.
   B. The level of scientific, mathematical, and technological activity and innovation reached or surpassed its Classical peak in the West.
      1. Practical concerns and the pursuit of "scientific" truths put a high premium on systematic experience and experimentation.
      2. The Humanists built an interconnected, international network of scholars sharing a common set of critical values and institutions.

II. Modern science can be new without any of its features being new.
   A. Modern science *was* new, and it *did* have a critical new feature.
      1. The new feature was method, the idea that achieving knowledge of nature was critically dependent on following a method.
      2. The discovery of "the" scientific method is said to have triggered the Scientific Revolution.
      3. The founders of modern science did share a concern with a problem posed by knowledge of nature that satisfied the Platonic-Aristotelian definition of *knowledge*: How can we have universal, necessary, and certain knowledge of nature if our only access to nature is experience—particular, concrete, continually changing?
      4. The founders of modern science used different methods to solve that problem.
      5. They also inherited a great deal from scientists of the 16th century.
   B. There seems to have been no such thing as "the" one scientific method, but method *was* a necessary condition of modern science.
      1. René Descartes and Francis Bacon proposed very different methods for discovering knowledge of nature.
      2. The most famous early modern scientists employed methods that were different from those of Descartes *and* of Bacon.
      3. To this day, there is no single specifiable method that a student can be taught that guarantees success in discovering new knowledge.
   C. Francis Bacon was an English lawyer, jurist, and educational reformer who is called the father of the experimental method.
      1. Bacon's interest in science was as a generator of the knowledge of nature that would translate into technological control over nature.
      2. In *The New Organon*, he argued that only a total reform of reason could lead to knowledge of nature.
      3. He cited four "idols" of the mind that must be overcome in order to reason effectively about nature because the greatest obstacle to achieving useful knowledge of nature was the mind itself.

4. Bacon proposed a strictly "mechanical" approach to the study of nature based on experimentation and *inductive* reasoning.
D. Bacon's only application of his method to a scientific problem then current was to the nature of heat.
   1. Heat was either motion of the small material particles of which large objects were composed, or an invisible fluid called *caloric*.
   2. Neither caloric nor the microscopic motions claimed to be heat can be observed directly.
   3. As with Copernicus, the object of scientific knowledge is an unexperienced reality that is the cause of experience.
   4. This is fully consistent with the definition of knowledge as universal, necessary, and certain.
   5. For Bacon, scientific knowledge accumulated over time and its ultimate test was the ability to predict and control experience.
   6. The founders of the Royal Society of London looked to Bacon as the guiding spirit of the new experimental philosophy of nature.
E. René Descartes has a greater claim than Bacon to be a father of modern science but of a very different sort of science.
   1. For Descartes, deductive reasoning was the only route to knowledge of nature, which he, too, claimed would give us useful power over nature.
   2. In his *Discourse on Method* and *Rules for the Direction of the Mind*, Descartes outlined a deductive methodology for science.
   3. Central to this were innate ideas and a mental faculty of intuition.
   4. He proposed a strictly mechanical philosophy of nature that evolved into materialistic determinism and prohibited any forces other than direct contact.
   5. Although this proved to be too limiting, it continued to be the metaphysical framework of modern science into the 19th century.
F. Descartes' deductive method had a strongly logical character, where Baconian induction had a strongly empirical character.

1. For Descartes, induction could never guarantee the truth of the universal generalizations that comprised knowledge.
2. Scientists needed to invent hypotheses that *could* be true and successfully deduce phenomena from them.
3. Isaac Newton opposed this methodology with its invention of hypotheses that "worked," arguing that the goal of science was to discover the "true causes" of natural phenomena.
4. Newton, like Galileo, practiced the brand of mathematical physics that Archimedes wrote about 1900 years earlier, in texts recovered and published by the Renaissance Humanists.
5. This is quite clear in Galileo's *Discourses on Two New Sciences*, his last book and his most important scientific work, though his *Dialogue Concerning the Two Chief World Systems* is better known.

Essential Reading:

Francis Bacon, *The New Organon*.

Richard S. Westfall, *The Construction of Modern Science*.

Questions to Consider:

1. It seems impossible to bridge the logical gulf between induction and deduction; what confidence can we have, then, in the truth of universal scientific theories and "laws" of nature?
2. Did Galileo really "see" moons orbiting Jupiter through his telescope, or did he see moving dots that he interpreted to be moons, and what difference does it make if the latter is true?

Lecture Fourteen
Algebra, Calculus, and Probability

Scope: European mathematics was in a pitiful state prior to the 16th century. Renaissance scholars' recovery and careful study of Greek and Graeco-Roman mathematical texts, together with the growing use of mathematics, stimulated original mathematical research. The 16th century was marked by the adoption, by way of Islamic texts, of algebra and theory of equations on a par with geometry-centered Greek mathematics; the creation of a primitive probability theory; and the introduction of a symbolic notation for arithmetic and algebra. Progress was explosive from the 17th century on. Descartes and others showed how to use algebraic equations to solve even the most complex geometric problems. Isaac Newton and Gottfried Leibniz invented the calculus, which proved to be immensely fertile both in "pure" and applied mathematics. With the calculus, for the first time, scientists had a means of exploring change. Meanwhile, probability theory expanded to incorporate the application of statistics to political and commercial decision-making.

Outline

I. The shift of European mathematics from geometry to algebra was epochal for the rise of modern science and for its impact on society.

  A. There is an interesting connection between the 17th-century focus on method that was the subject of the preceding lecture and the shift from geometry to algebra.
   1. Bacon, Descartes, Galileo and Newton promoted different versions of "the" scientific method.
   2. Bacon's was inductive, Descartes' was deductive while Galileo and Newton adopted a broadly Archimedean methodology that was mathematical, experimental, and deductive.
   3. The common denominator was they were all radically impersonal and value neutral.

    4. All that matters in the modern-science style of reasoning is the methodology employed, the results, and the criteria for evaluating the results, for example, confirmatory experiments.
    5. This impersonalism, this objectivity, and the focus on logic in the study of nature are reminiscent of mathematics.
B. The mathematics of the early 16th century was overwhelmingly Greek, but by the end of the century, it was very different.
    1. Greek mathematics was almost wholly geometric, including trigonometry introduced to deal with problems in astronomical theory.
    2. Especially during the Graeco-Roman period, Greek mathematicians explored number theory and equations, but overwhelmingly, the preference was for transforming all mathematical problems not obviously arithmetic into geometric terms.
    3. This preference seems related to the identification of geometric reasoning with deductive reasoning and its guarantee of (logical) truth.
    4. Islamic mathematicians mastered and extended Greek geometry and trigonometry, but they also cultivated "algebra." Al-Khwarizmi in the 9th century and abu-Kamil in the early 10th century became major influences when their books were translated into Latin during the Renaissance. The very term *algebra* derives from an Arabic word (*al-jabr*) used by al-Khwarizmi.
    5. Islamic mathematicians also transmitted to the West the Indian number symbols, including the use of zero, that became widely adopted only in the 16th century, finally displacing Roman numerals.
C. This great legacy converged in the 16th century on a peculiarly receptive western Europe that responded enthusiastically and creatively.
    1. The wholesale adoption of the new number symbols was of fundamental importance, especially when complemented by the invention of a symbolic language for algebraic expressions.
    2. This symbolic language played an important role in the sudden increase in mathematical creativity, but it was a product of an antecedent "desire" to pursue mathematics.

3. Important new results were achieved, among them, a general solution of the cubic equation and the creation of probability theory, both by Italian mathematicians (though Omar Khayyam seems clearly to have anticipated Tartaglia's solution of cubic equations by about 400 years); advances in trigonometry crucial to astronomical theory; mapmaking and surveying; and the invention of complex numbers and logarithms.
   4. Tartaglia, a nickname for Niccolò Fontana, and Ludovico Ferrari, a one-time servant to the physician-mathematician Jerome Cardan (Girolamo Cardano), solved the cubic and quartic equations, and Cardan introduced—among other innovations—probability theory, in a book not published for 150 years.
D. René Descartes was a father of modern philosophy and modern science—championing a deductive methodology, a mechanical philosophy of nature, and the mathematization of physics—and a major contributor to modern mathematics.
   1. Descartes paved the way to analytic geometry, to equation-based geometry, by freely introducing *arithmetical terms* into the solution of geometric problems.
   2. In his book titled *Geometry*, he states that every geometric problem can be solved algebraically, whether the problem is mathematical or "scientific."
   3. Even as he equated matter and space in his theory of nature, Descartes reinforced the growing focus on algebra, not only in mathematics, but in physics, as well.
   4. Descartes' *Geometry* is one of the earliest mathematical texts employing the full symbolic notation we use today, and thus, it is quite readable.
E. Algebra steadily displaced geometry in European mathematics (though geometry underwent a revival in the 19th and 20th centuries in a much more abstract, algebraic form), but it encountered strong resistance.
   1. Galileo did not once use the term *algebra* in his *Two New Sciences*, and Newton employed only geometric proofs in his *Principia*!

2. The resistance seems to reflect a lack of confidence in the deductive character of algebraic solutions compared to the clearly deductive character of geometric proofs.
3. Using algebra solves the problem but not in a logically compelling way, almost as if a trick had been employed.

II. The calculus may have been the most creative and the most consequential mathematical invention of the 17th century, followed by the invention of probability theory.
   A. The origins of the calculus lie in technical mathematical problems, but its consequence was to give unprecedented power to scientists modeling natural phenomena.
      1. There are two "sides" to the calculus: the integral side, rooted in attempts to calculate areas and volumes, and the differential side, rooted in attempts to finding the tangent to a curve at any given point on the curve.
      2. Newton and Gottfried Wilhelm Leibniz independently invented the calculus at very nearly the same time.
      3. Subsequently, they engaged in one of the bitterest priority disputes in science, with Newton behaving particularly badly.
      4. The "genius" of the differential calculus, and its power for science, lies in the ability it confers to measure change as it calculates the direction of the tangent to a curve at any point on the curve.
      5. Geometry is static and timeless, but an algebra enriched by calculus is dynamic.
      6. With the calculus, one can give a mathematical description of processes as they change in time, using the differential calculus to calculate the direction of change and the integral calculus to calculate the cumulative consequences of change.
      7. Furthermore, the calculus was readily extended to all processes in the life and social sciences, as well as the physical sciences.
   B. Plato and Aristotle, building on Parmenides and Pythagoras, had defined knowledge as universal, necessary, and certain—though Sophists and later the Skeptics rejected this definition.
      1. If knowledge is universal, necessary, and certain, then a *science* of probabilities seems oxymoronic.

2. Probability theory, in this context, is an idea that had to be invented.
3. Though there were predecessors, perhaps the founding text of modern probability theory was Jacob Bernoulli's *The Art of Conjecturing*.
4. What distinguished Bernoulli's book was his claim that probability theory was the basis for rational, effective, real-world policy- and decision-making for governments and individuals.
5. His idea for a *science* of probabilities was applied in the 18th and 19th centuries, and it became a major feature of 20th-century social science.

Essential Reading:

Jacob Bernoulli, *The Art of Conjecturing*, translated by Edith Dudley Sylla.

Ivor Grattan-Guiness, *The Norton History of the Mathematical Sciences*.

Questions to Consider:

1. What is it about symbolic notations in mathematics that makes them so fertile?
2. Hindu-Arabic numerals first appeared in Europe in the 10th century; why were they not adopted until the 16th century?

Lecture Fifteen
Conservation and Symmetry

Scope: The purpose of science is to give us knowledge of nature derived from, and validated by, experience. The feature of experience that above all demands explanation is change, but knowledge purports to be changeless. How can we have a changeless account of the continually changing? This problem had been raised by the first Greek philosophers, who offered two responses that have survived to the present. One, which evolved into modern atomic theory, supposed that the ultimately real was composed of changeless, elementary substances whose properties in various combinations generated the rich diversity of experience. The other supposed that change was elementary, that reality was a web of processes, but all change expressed changeless patterns of change. This line evolved into modern field theories of matter and energy. Common to both lines are assumed principles of conservation and symmetry that underlie change and make scientific explanation possible.

Outline

I. *The* fundamental fact of experience is change, which we experience as relentless and universal—everything in experience changes all the time—thus, *the* fundamental task of natural science is to explain change.

    A. At the very beginning of Western philosophy, the problematic character of change was identified as a core issue.

        1. Parmenides wrote a book-length poem arguing that change could not possibly be real because being and not-being are mutually exclusive categories, yet change implies that something that is, is not and that something that is not, is.

        2. Parmenides went so far as to argue that it was not even possible to think about not-being, let alone reason logically about it.

        3. Though flatly contradicting everyday experience, Parmenides's critique of change provoked a refined materialism as a response.

4. On this view, changeless elementary forms of matter are the ultimate constituents of reality.
5. One version of this materialism was Empedocles's theory of four elementary forms of matter: earth, air, fire, and water.
6. Another was atomism, especially as developed by Epicurus and Lucretius.

B. It was easy to mock or dismiss Parmenides's conclusions but not the point he raised about the at least apparent non-rationality of change.
1. Heraclitus made the reality of change the cornerstone of his philosophy and the philosophy of nature: "Everything changes; no *thing* remains [the same]."
2. But if everything changes all the time, then how can we make sense of experience, let alone have *knowledge* of it?
3. Heraclitus's answer was that although change is real, it is rule-governed: Nature is a collection of processes, each of which has a distinctive *logos*, or rule/law, that endures over time and gives that process its identity.
4. For Heraclitus, the goal of knowledge of nature, of what we mean by *science*, is to discover lawful patterns of change, and this makes the calculus a particularly powerful analytical tool.

II. As modern science took shape in the 17th century, two ideas emerged as fundamental to its practice: conservation and symmetry.

A. Symmetry functioned implicitly until the 19th century, but conservation was explicit from the start.
1. Change is a fact of experience and so is orderliness.
2. Science is not needed for people to recognize patterns in everyday natural phenomena.
3. Seasons change in regular ways; stars and planets move in recurring patterns; plants and animals reproduce predictably.

B. What experience does not reveal is a necessary order to natural phenomena, an absolute order to nature.
1. Science enters with the proposition that there is a necessary order to experience and that it is the task of science to expose this order.
2. Science explains changing natural phenomena by invoking something that is unchanging.

3. If everything changes all the time, patternlessly, then explanation is not possible.
4. Either some explanatorily fundamental things do not change or some patterns do not change.

C. Determinism is inescapable if explanation has to satisfy an idea of knowledge based on deductive reasoning.
1. The introduction of the falling-weight–driven mechanical clock quickly was followed by the metaphor that the Universe was a clockwork mechanism.
2. The clock measures time, of course, which is always changing, yet the clock has a deterministic structure to its design and operation.
3. Every gear-train mechanism, of which the mechanical clock is one type, has a deterministic character by design.
4. To say that the Universe is a clockwork mechanism is to say that nature, the Universe, is deterministic.

D. We noted earlier the idea that nature is a closed system; this idea contains the seed of the idea of conservation.
1. Given deterministic change within a closed system, one of two things must happen.
2. Either the system steadily "winds down," as mechanical clocks do, or it is stable, in spite of continual change.
3. Natural patterns seem very stable, with no sign of winding down; thus, there must be *something* invariant *within* change, some aspect of change must be conserved that keeps the Universe from winding down.

E. In the late medieval and Renaissance periods, this *something* was motion.
1. The conservation of motion was, until the mid-17th century, the first and the bedrock law of conservation in science.
2. Mechanics, that branch of physics that describes the laws of motion of material objects, began to be practiced in the 16th century.
3. Galileo was trained in an Italian school of mechanics that succeeded in identifying laws of motion for moving objects.
4. The conservation of motion was a necessary condition of the very possibility of these laws.

III. The second great conservation law in modern science was the conservation of matter.
   A. The conservation of matter and the conservation of motion play complementary roles, and both turned out to be wrongly conceived.
      1. Like the conservation of motion, the conservation of matter is implicit in the late-medieval naturalistic determinism of the closed-system clockwork-mechanism metaphor.
      2. Although it was recognized by the 16th century, the conservation of matter did not become an explicit principle of nature until the late 18th century and, in particular, in the work of Antoine Lavoisier.
      3. Lavoisier's oxygen-based theory of combustion inaugurated modern scientific chemistry and the atomic theory of matter as built up out of elementary forms of matter.
   B. Descartes' mechanical philosophy of nature, influential from the mid-17th through the 19th centuries, is based on these two conservation laws.
      1. Nature, for Descartes, is matter in motion, and that's it!
      2. All natural phenomena, with only one exception, are to be explained in terms of the distinctive motions of distinctive complexes of matter.
      3. The exception for Descartes, but not for later materialists, was the human mind and its free will.
      4. Descartes' materialism stands in a tradition of those ancient Greek philosophers called *materialistic monists*, who believed that there was one generic type of "stuff" out of which all material objects were composed.
   C. Descartes' explanation of free will revealed flaws in his conception of the conservation of motion.
      1. Descartes' anatomical studies led him to propose that the pineal gland was a kind of switchboard directing nervous fluid throughout the body.
      2. Descartes defined motion as mass multiplied by speed, arguing that the direction of the nervous fluid could change without violating the conservation of its motion.
      3. Around 1650, Christiaan Huygens showed that a force is required to change the direction of a moving object, even if the speed and mass are constant.

       4. This implied that the "correct" definition of the motion that was conserved was mass multiplied by velocity, a quantity we call *momentum*.

       5. The conservation of momentum, linear and angular, became an inviolable principle of nature/reality, and it remains so to the present day.

D. As modern science evolved, the "thing" or quantity conserved was not obvious.

       1. What remained obvious and inescapable was that some fundamental things must be conserved in order for scientific explanation to be possible.

       2. As we will see, the creation of thermodynamics in the 19th century created energy as a new elementary feature of reality and a new law: that energy was conserved.

       3. At the end of the 17th century, Newton and Leibniz argued over the significance of the quantity mass times velocity squared.

       4. Newton dismissed it as a mere number, but Leibniz claimed that he had discovered a new conserved feature of nature that he called *vis viva*, or "living force."

       5. In the 19th century, *vis viva* became *kinetic energy*, which is indeed conserved, but this dispute reflected a much deeper dispute over how deterministic the Universe was and what role this left for God.

       6. In 1905, just 50 years after the law of the conservation of energy was proclaimed as a great triumph of science, Einstein overturned it with his discovery that matter and energy were interconvertible.

E. In the 19th and 20th centuries, the idea of conservation as an elementary feature of nature was joined by the ideas of invariance and symmetry.

       1. In effect, invariance and symmetry are mathematical versions of conservation of physical quantities.

       2. Both ideas reflect the intensification of the connection between mathematics and descriptions of natural phenomena.

       3. By the end of the 19th century, for many physicists, the mathematics *was* the phenomenon, such that a mathematical invariance or symmetry implied the conservation of a physical quantity.

Essential Reading:

Mary Hesse, *Forces and Fields*.

Peter Pesic, *Abel's Proof*.

Questions to Consider:

1. Are the Parmenidean and Heraclitean accounts of reality exclusive of each other, as they have been taken to be historically, or could they be complementary?
2. How do scientists know that constants are truly constant "forever"?

Lecture Sixteen
Instruments as Extensions of the Mind

Scope: Ordinarily, we think of instruments as extending the senses, analogous to machinery extending the muscles, but this is misleading. It is a fundamental premise of modern science that the reality it seeks to describe as the *cause* of sense experience is fundamentally *different* from sense experience. The microscope and the telescope, among the earliest instruments associated with modern science, seem to be extensions of our visual sense, but again, this is misleading. Typically, we cannot verify independently, for example, by visual inspection, what the microscope and the telescope show us. This is also true of such simple instruments as the air pump, barometer, and thermometer and, of course, of such complex instruments as electron and scanning tunneling microscopes, DNA sequencing machines, high-energy particle accelerators, and neutrino and gravity telescopes. Modern science is a web of concepts, and the instruments it employs are conceptual: What they tell us about nature is a matter of interpretation, not observation.

Outline

I. Scientific instruments have not received nearly as much attention from scholars as scientific ideas, perhaps because they are associated with know-how rather than with knowledge.

    A. It is a commonplace to refer to instruments as extensions of the senses, but in the overwhelming majority of cases, they are extensions of the mind.

        1. Some simple instruments, for example, magnifying lenses and spectacles, *are* extensions of our senses, as basic machines are extensions of our physical strength/power.

        2. We tend to assume that this is true for instruments generally, but this is not correct.

        3. Even relatively simple instruments, such as telescopes and microscopes, are really extensions of the mind, of how we conceptualize and explain experience.

4. When the results obtained by using an instrument cannot be verified independently by the senses, then the instrument is the embodiment of ideas.

B. The first modern scientific instruments are, not surprisingly, 17th-century phenomena, and they are clearly extensions of the mind, if not projections of the mind.
1. As extensions/projections of the mind, the telescope, microscope, air pump, and barometer tell us at least as much about how people conceptualize nature as they do about nature.
2. Consider Galileo's shift from using the telescope to see distant objects on Earth and in the sky. The former can be verified by inspection but not the latter: for example, his claim that there are moons orbiting Jupiter.
3. The same is true of the microscope and its "revelation" that there are tiny, invisible "animals" in our drinking water.

C. As extensions of the mind rather than the senses, what instruments "reveal" about nature cannot be verified by direct inspection: We need to "trust" the instrument and the ideas/theory on which its construction is based.
1. Trusting instruments means trusting the ideas/theories in accordance with which the instrument is constructed.
2. How do we know whether an instrument whose results we cannot verify independently is revealing truths about nature or "truths" about the (mal)functioning of the instrument?
3. In fact, in the 17th and 18th centuries, both microscopes and telescopes suffered from optical aberrations that produced misleading, blurry, false-colored images.
4. These aberrations became worse as magnification increased, and everyone wanted higher magnification, especially for the microscope.
5. Techniques for making color-corrected microscopes were invented in the late 18th century but didn't become available until the 1820s.
6. Simple microscopes of the late 17th century had allowed Anton van Leeuwenhoek and Robert Hooke to make wonderful "simple" discoveries, but 19th-century microscopes allowed biologists to study the internal structure of cells and their nuclei.

II. The telescope and microscope are the best known of the instruments used by the 17th-century founders of modern science, but they are not the only ones.

   A. The air pump, barometer, and thermometer are also important 17th-century instruments and also embody ideas.
      1. The barometer was invented for the explicit purpose of testing an idea: Was there a vacuum?
      2. The prevailing view, inherited from Aristotle and defended by Descartes, among many others, was that a vacuum was impossible.
      3. Atomists, among them Galileo, defended the existence of "local" vacua between atoms, and some accepted Epicurus's version of atomism, in which atoms and "the void" are the ultimate reality.

   B. For both religious and scientific reasons, the reality of the vacuum was a controversial issue in early modern science.
      1. The Catholic Church was hostile to Epicurus's atomic philosophy because it was explicitly anticlerical and denied the soul or life after death.
      2. Blaise Pascal, a brilliant, young French natural philosopher and mathematician, made early barometers and believed that the empty space at the closed end of the column of water or mercury was a vacuum.
      3. One of the experiments he proposed and arranged to be carried out was to carry a barometer up a mountain, observing any changes in the height of the supported column of mercury.
      4. His conclusion was that the atmosphere, air, possessed weight and exerted a mechanical pressure, just as a liquid does.
      5. He also claimed that this proved that the empty space in the barometer tube was a true vacuum, but many others disagreed, interpreting the results of this experiment in a way that was consistent with the denial of a vacuum.

   C. How do we know what an instrument, especially a new kind of instrument, is telling us about nature?
      1. If a thermometer gives us a temperature reading that is not manifestly ridiculous, we assume that it is operating correctly.
      2. The air pump, developed into a research instrument by Robert Boyle and his then-assistant Robert Hooke, was particularly problematic in this regard.

3. Boyle and Hooke used their pump, which was fragile, finicky, and malfunctioned often, to proclaim new laws of nature relating to the pressure, volume, and temperature of air.
4. Thomas Hobbes, a Cartesian-style rationalist philosopher, famous for his social and political philosophy rather than his attempts at natural philosophy and mathematics, criticized Boyle's results.

D. Following Descartes, Hobbes was deeply suspicious of complicated experiments.
1. Hobbes was not prepared to trust that a newfangled machine could reveal nature's secrets, and he argued forcefully that experiments were open to multiple interpretations: Experience with the barometer showed that.
2. This illustrates very well that instruments are extensions of the mind.
3. Hobbes attacked the air pump as producing the results obtained by Boyle and Hooke, not simply revealing facts about nature.
4. We need to keep this in mind with respect to complex instruments of today, such as DNA sequencing machines and computer simulations.
5. Even a simple electronic calculator is almost invariably accepted on trust, but complex simulations involving programs produced by a large team of researchers are especially problematic.
6. Particle accelerators produce "real" physical events, typically, millions of collisions between subatomic particles, but those events are recorded and analyzed by extremely complicated instruments and computer programs.
7. The "discovery" of the sixth quark, called *top*, in 1995, involved identifying half a dozen "sightings" of the top quark among 15 million events!
8. Consider neutrino telescopes, buried deep underground, "observing" a minute fraction of a fraction of neutrinos passing through the Earth or, perhaps, just randomly "firing."
9. Only the theories used to build a device and interpret its output—ideas, in short—can distinguish between static and meaningful signals.

III. There would seem to be a logical circularity in claiming that instruments that embody ideas and theories reveal a reality that is independent of experience.

  A. The study of scientific instruments with respect to their role in theories and explanations raises questions about the nature of scientific knowledge.
  1. The Sophists, recall, argued that knowledge was about experience and was a form of what we would call high-probability belief about experience validated by experience.
  2. It was Plato and Aristotle who championed the rival view that knowledge was universal, necessary, and certain truth about a reality independent of experience.
  3. The role of instruments in the logic of theory formation and explanation seems closer to the Sophist view than the Platonic-Aristotelian.

  B. It is also true that instruments designed for one purpose sometimes give completely unintended results, leading to new theories or revisions of existing ones.
  1. Instruments invented to detect the *aether* in late-19th-century physics gave results that led to the conclusion that the aether did not exist.
  2. Radio frequency radiation from outer space was discovered by telephone company engineers studying static on long-distance lines.
  3. The discovery, dating, and analysis of fossils, for example, using radioactive dating and DNA testing, has led to multiple theories and changes in theories.

  C. The relationship between physical instruments and theories and ideas is complex and reciprocal.
  1. It is simplistic to say that instruments are ideas only, incapable of telling us something we were not already committed to.
  2. It is also simplistic to dismiss scientific explanation as logically circular because of this reciprocity.
  3. It is important to realize that instruments are extensions of the mind, not the senses, and as such, the relationship between instruments and theories demands much more attention than has been given to it.
  4. This has become increasingly important as the most sophisticated instruments require our most sophisticated

theories for their design and for the interpretation of their results.

Essential Reading:

Davis Baird, *Thing Knowledge: A Philosophy of Scientific Instruments*.

Steven Shapin and Simon Schaffer, *Leviathan and the Air Pump*.

Questions to Consider:

1. If instruments embody ideas, can observation be independent of the theories that observation is supposed to confirm or disconfirm?
2. As instruments become more complex and experiments more expensive, what happens to replication by others as the traditional test of the validity of scientific knowledge claims?

Lecture Seventeen
Time, Change, and Novelty

Scope: Time is not a fundamental feature of reality, either for modern science or for the philosophical tradition whose definition of *knowledge* science assimilated. If the ultimately real—whether God, elementary substances, or laws of change—is changeless, it is also timeless. This timelessness of the real appears in modern science as the reversibility of time in the equations of mathematical physics. The irreversibility of time *as experienced by us* is merely a fact about us, reflecting the limitations of our ability to experience reality. Although the 19th century began with a declaration of the ultimate insignificance of time, developments in the course of the century challenged this. The concept of energy, the new science of thermodynamics, and the theory of evolution implied that time truly was irreversible and that it was the dimension in which unpredictable novelty emerged. The debate over the nature of time continues to rage even today.

Outline

I. How to explain change has been the fundamental challenge for knowledge of nature since Parmenides and Heraclitus, and because time is the measure of change, it is pivotal to scientific explanation.

    A. That instruments are extensions of the mind makes an issue out of the relationships among instruments, ideas, and tools.
        1. As physical devices, scientific instruments seem very different from what we commonly mean by *tools*.
        2. Consider the obvious difference between the Hubble Space Telescope and a screwdriver.
        3. Conceptually, however, even a tool as simple as a screwdriver or a hammer has a great deal in common with complex instruments: Both embody systemic ideas.

    B. Although it seems odd to say so, scientific ideas and concepts—ideas become concepts when defined in more specific ways—are tools.
        1. It is by means of concepts that scientists analyze phenomena and construct explanations.

2. How the concepts scientists use are defined is central to doing science, and for physics, the most fundamental concepts include space and time.
3. Keep in mind that we are talking about ideas as they function in scientific explanations and that *all* scientific ideas change over time—even the idea of time itself!

C. Perhaps the most famous philosophical analysis of the idea of time is in Augustine's *Confessions*.
1. Augustine concluded that *time* was a name for a mental phenomenon, not a feature of reality.
2. Time is coordinate with, and the measure of, change; the changeless is timeless.
3. But if we want to know what time is, we are puzzled because time's manifestation in consciousness lacks being!
4. The past no longer is; the future is not yet; and the present is a timeless instant (so Augustine claimed) between past and future.
5. Augustine concludes that time is a feature of the mind, not reality.

D. Newton's masterwork, the *Mathematical Principles of Natural Philosophy*, founded the modern science of mechanics; he begins with definitions of space, time, and motion.
1. Newton notes that these ideas are familiar to everyone, but this simply means that, as Francis Bacon warned, we have preconceptions that need to be clarified.
2. Before he can write equations describing the motions of material objects under various forces, he needs precise definitions of these ideas.
3. It is especially important for Newton to distinguish relative and absolute time, space, and motion.
4. Newton defined time, as well as space and motion, as having an absolute character.
5. Time exists independently of anything that happens *in* time.

E. Newton defined time as an absolute clock, "ticking" away at an absolute, uniform rate.
1. This idea of time, and Newton's definition of space, as well, remained the cornerstone of mechanics and, hence, of physics for 200 years.

2. There was an implicit problem in how these definitions can be correlated with the actual measurement of time and space.
3. This problem became explicit in the late 19th century, and in the early 20th century, Einstein's relativity theories forced redefinitions of time and space that abandoned their absolute character and made them into relationships.

II. There is a philosophical aspect to the idea of time that underlies and even haunts its scientific definition.
    A. Those Greek philosophers for whom knowledge was universal, necessary, and certain valued the timeless over the temporal.
        1. The centrality of deductive reasoning to the idea of knowledge as formulated by Parmenides, Plato, and others implies that knowledge, truth, and reality are timeless.
        2. The conclusion of a deductive argument is already *contained* in the premises: It is only "accidental" that we human thinkers have to draw it out for examination in an explicit conclusion.
        3. Transposing this deduction-based philosophical idea of knowledge to knowledge of nature implies that effects already exist *in* their causes; thus, natural time, too, is a mere "accident."
    B. In fact, modern science assimilated just this set of ideas.
        1. The mathematico-deductive methodology of Archimedes was, as we have seen, adopted by Galileo and Newton, and it became the "gold standard" of scientific theory construction.
        2. Scientific theories take exactly the form of deductive arguments, from which phenomena to be explained are deduced.
        3. But that means, ultimately, that time is not real, as Augustine concluded on philosophical grounds.
        4. *We* need to draw conclusions because our minds are limited in power, but an unlimited mind would see everything in an instant.
        5. In the early 17th century, Galileo said this in his *Dialogue*, and later in that century, Leibniz wrote that the present is "pregnant' with the future, as the past had been "pregnant" with the present.
    C. This is the essence of determinism, which is timeless.

1. Leibniz's dictum found literal expression in contemporary 18th-century embryological theories of *preformationism*.
2. Embryos were fully preformed neonates, minutely folded inside the egg or the sperm, and unfolded only during gestation.
3. In physics, the ultimate timelessness of reality was reflected in the *reversibility* of equations in mechanics: Time can go either way.
4. At the turn of the 19th century, Pierre-Simon Laplace proclaimed the indifference of the material world to time in the context of his proclamation of the exhaustive character (in principle) of the materialistic determinism of mathematical physics.

III. The prevailing cultural and scientific conception of time changed in the 19th century, even as Laplace's formulation became canonical.
   A. The idea of progress entails an asymmetric conception of time, as does the idea of historicity.
      1. The future orientation of Renaissance believers in progress makes time not just the dimension of change but change for the better.
      2. History, as an account of change over time in human affairs, achieves a new stature.
      3. The Renaissance Humanists, contra Aristotle, made history a "science," that is, a body of true knowledge and even the model for knowledge, in spite of its particularity and contingency.
      4. Through their studies of the past, the Humanists made intellectuals in the early modern period aware of the historicity of language, law, and institutions, that these changed *irreversibly* over time.
      5. The idea of progress makes time into the dimension of human hope and even salvation.
   B. In the 18th century, the Humanist sensitivity to history and to historicity expanded its scope.
      1. Giambattista Vico's *New Science* proposed that human consciousness evolved, that *how* we thought changed over time.

    2. Eighteenth-century theories of the origin of the Solar System and the Earth attributed change to them, contrary to Aristotle's claim that they had eternally been the same.
    3. At the turn of the 19th century, Gottfried Herder argued for the historicity of human culture in all its forms.
    4. Early in the 19th century, Hegel attempted a new form of deductive logic in which time played a significant role—a dynamic determinism as opposed to simple preformationism—which became the "inspiration" for Karl Marx's *scientific materialism*.

C. Time, as defined explicitly by Newton and implicitly in the idea of scientific knowledge, became an issue in science in the 19th century.
    1. The idea of energy that blossomed into the new theory of heat called *thermodynamics* led to a new "law" of nature that implied that time was *irreversible*, that it could flow in only one direction.
    2. This flatly contradicted the *reversibility* of time in mechanics.
    3. The Darwin-Wallace theory of evolution, contemporary with the invention of thermodynamics, also implies that time is a dimension of the emergence of novelty, not simply the unfolding of what is already there in germ.
    4. Such developments created a tension in science between those committed to a deductive-deterministic conception of knowledge and reality and those for whom new theories implied that knowledge and reality both had a statistical-probabilistic character.

Essential Reading:

Ilya Prigogine, *Order Out of Chaos*.

Martin J. S. Rudwick, *The Meaning of Fossils*.

Questions to Consider:

1. What is revealed about Western intellectualism by the fact that so many of our greatest philosophical and scientific thinkers believed that time was an illusion?

2. Is a deterministic account of nature compatible with a conception of time as irreversible?

Lecture Eighteen
The Atomic Theory of Matter

Scope: The Parmenidean view that the ultimately real was changeless found expression even in antiquity in atomic theories of matter, which became prominent again in the late Renaissance. Newton's atomism explicitly proclaimed the timeless permanence of atoms and their properties, but atomism became a functional scientific theory only in the early 19th century, when John Dalton argued that an atomic theory of matter explained chemical reactions and some puzzling physical phenomena. Chemists quickly took up the theory and used it with great success, though many physicists rejected the reality of physical atoms until the early 20th century. While this debate raged, the generic approach of explaining phenomena by postulating fundamental entities, "atoms" broadly speaking, whose properties cause the phenomenon to be explained, flourished. Instances of this way of thinking include the cell theory of life, the germ theory of disease, the gene theory of inheritance, and the quantum theory of electromagnetic energy.

Outline

I. That matter is atomic in nature was an idea, hence an invention, not a discovery.

    A. As we saw with time, there is an important difference between an idea and a scientific idea.

        1. Augustine's philosophical idea of time is far more sophisticated than common usage, but it is not a scientific idea.

        2. In Newton's *Principia*, we saw that a scientific idea is one whose definition is adapted to the context of a scientific explanation or theory.

        3. The fact that scientific ideas change, as did the idea of time in the 19th and 20th centuries, reveals the contingency of scientific definitions and, hence, ideas.

    B. The atomic idea began in antiquity and, until 1806, was a metaphysical speculation.

1. Greek atomism was a response to Parmenides's logic-based critique of the reality of change.
2. Atoms explain change as a rearrangement of things that are themselves changeless.
3. Anaxagoras postulated a variety of atoms as broad as the variety of things in nature.
4. Epicurus's atomism claimed that natural objects were combinations of atoms that possessed only size, shape, weight, and motion.

C. Epicurus's atomic theory was explicitly proposed to support a moral philosophy hostile to organized religion.
  1. Epicurus taught that life was an irreducibly material phenomenon that ended with death and should be lived as such.
  2. Reality was a vast but finite "rain" of atoms falling eternally in infinite space, randomly swerving, colliding, and forming more-or-less stable agglomerations and hierarchies of agglomerations that dissolve and form new agglomerations.
  3. Epicurus's theory was disseminated in the Roman era by Skeptical philosophers and Lucretius, among others, and was anathema to Christian authorities.

D. Atomism entered modern science through Humanist texts and in spite of Church opposition to the dissemination of Epicurean ideas.
  1. In the 16th and early 17th centuries, natural philosophers concerned with explaining and describing the motion of matter saw the atomic theory of matter and the idea of empty space as useful.
  2. Galileo defended atomism.
  3. Robert Boyle, whose (Protestant) religious orthodoxy was beyond question, used an atomic theory to explain chemical reactions, emphasizing size, shape, and motion as the elementary characteristics of atoms.
  4. Pierre Gassendi was a prominent contemporary atomist and a probabilist, opposing both to Descartes' necessitarian nature philosophy.
  5. Descartes rejected both atomism and the possibility of empty space, making nature a matter-filled plenum: For anything to move, everything must move!

E. Newton wove the atomic idea of matter into his theories of nature.

1. In his *Opticks*, Newton speculated that God had created matter in the form of minute spheres, indestructible and possessing innate properties, especially forces of selective attraction and repulsion.
2. His theory of light was a "corpuscular," or atomic theory, and although the *Principia* is not explicitly atomic, the spread of Newtonian mechanics in the 18th century came at the expense of Cartesian anti-atomism and disseminated at least the idea of atomism.
3. Leibniz elaborated a complex metaphysical system in which the ultimate reality is a vast array of immaterial "atoms" called *monads*, each the center of a unique set of forces expressive of a unique manifestation of being.
4. Leibniz's physics and mathematics were fundamental to Continental science in the 18th century, as were Newton's, and both Newtonian and Leibnizian metaphysics were widely disseminated.
5. Roger Boscovich's 1758 *Theory of Natural Philosophy* offered a Newtonian-Leibnizian theory in which matter is composed of "atomic" point particles that are the centers of innate forces acting at a distance.

II. At the start of the 19th century, then, the atomic idea was familiar in philosophical and scientific circles, but scientific atomism is said to have begun with John Dalton.

   A. Dalton's *New System of Chemical Philosophy* first appeared in 1807, though he started developing his atomic theory of matter by 1803.
      1. Dalton was a natural philosopher with a wide range of interests, but he had a special interest in meteorology and atmospheric pressure.
      2. Under the influence especially of Lavoisier's chemical revolution of the 1780s, Dalton built his explanation around the new definition of an element and an unshakable conviction that elements combine only in fixed proportions to form compounds.
   B. While the bulk of Dalton's research lay in what we would call physics, his atomic theory of matter was of immediate, direct relevance to chemistry.

1. Many of the leading chemists rejected Dalton's fixation on laws of fixed proportions for chemical reactions, but Dalton persevered.
2. Making the "weight" of hydrogen 1, he calculated the relative weights of many other elements, assuming that atoms differed only in weight, size, and innate properties, à la Newton.
3. Throughout the 19th century, chemists and physicists were comfortable with using the atomic theory of matter, but most physicists did not accept that matter really was atomic.

C. Acceptance of the reality of atoms grew steadily in chemistry from the mid-century but in physics only in the decade after 1900.
1. Chemical research, leading to a growing power to control increasingly complex reactions, reinforced the idea that elements combine in fixed, though multiple, proportions.
2. Concurrently, chemists discovered the significance of molecular structure in giving compounds with the same number of the same atoms different physical properties.
3. The ability to solve practical chemical problems with the atomic theory was reinforced by the kinetic theory of gases.
4. The combination of these developments should have clinched the reality of atoms even for physicists, but this was not the case; Ernst Mach remained a violent opponent of real atoms into the 20th century.
5. Boltzmann became convinced of the reality of atoms, but Maxwell waffled, although the statistical interpretation of thermodynamics and the kinetic theory of gases strongly suggested their reality.

III. Historically, the idea of the atom is of a simple, indivisible, and indestructible unit, but atoms became complex, divisible, and destructible.

A. The new understanding of the atom and its technological exploitation began with the discovery that atoms have parts and are mutable.
1. In 1897, J. J. Thompson announced the existence of lightweight, negatively charged particles *within* the body of the atom.

- 2. This required a radical redefinition of *atom*, which from the Greeks through Newton and Dalton to 1896, had been an irreducible, solid unit.
- 3. In 1911, Ernest Rutherford announced that the atom was overwhelmingly empty space!
- 4. In the 1930s, the nucleus was determined to have an internal structure and also to be largely empty.

B. By WWII, the atom was composed of protons, neutrons and electrons, with hints of other particles.
- 1. By the 1960s, physicists had identified over two hundred "elementary" subatomic particles, leading to a new theory of matter in which protons and neutrons are composed of quarks bound together by gluons.
- 2. At every stage in the development of modern physical science, the atom, purportedly *the* fundamental building block of reality, was redefined to fit the prevailing context of theory and explanation.
- 3. In the context of quantum theory, atoms even acquired qualities of waves, as we will discuss in a later lecture.

Essential Reading:

Carlo Cercignani, *Ludwig Boltzmann: The Man Who Trusted Atoms*.

Bernard Pullman, *The Atom in the History of Human Thought*.

Questions to Consider:

1. What are we to make of the fact that the more the atom was accepted as physically real, the emptier it became of anything solid?
2. Must atoms be ultimate units of solidity with fixed properties, or can they be relatively stable processes?

Lecture Nineteen
The Cell Theory of Life

Scope: What is life? This became a scientific question—to be resolved within a theory of nature—in the 18th century. The Cartesian school held that all life processes in their various manifestations were purely mechanical phenomena, consistent with an exhaustive materialistic determinism. Against this, it was argued that life phenomena were not reducible to matter in motion only: They were governed by forces and laws unique to the domain of living things. With the advent of microscopes with high-power lenses corrected for distortion, it was argued that there was a common building block to all forms of life: the cell. From about 1835 to the end of the century, progressively more sophisticated observational studies advanced the cell theory of life, which together with the germ theory of disease and the gene theory of inheritance, evolved into a mutually reinforcing web of correlated ideas.

Outline

I. The atomic theory of matter is an instance of a particular style of thinking in science, a "Parmenidean" approach to explaining natural phenomena.
   A. Other scientific theories exhibit this same style of thinking.
      1. The cell theory of life, the germ theory of disease, and the gene theory of inheritance are examples.
      2. What all have in common is the postulation of some elementary unit with fixed properties that produces the phenomena to be explained.
      3. The cell, with its internal structures and processes, is the basis for explaining all life phenomena, which we cluster under the rubric *biology*.
      4. This style of thinking is deeply ingrained in Western culture and at every level of social, political, and scientific thinking.
      5. It contrasts sharply with the process style of thinking characteristic of science since the mid-19th century.
   B. Biology, as a unified science of life, is a product of the turn of the 19th century.

1. Botany, zoology, and medicine were all clearly "scientific" by the 18th century.
2. Throughout that century, rival interpretations of life were pursued, one mechanistic and the other vitalistic.
3. Around 1800, Jean-Baptiste Lamarck coined the term *biology*, consistent with his evolutionary theory of life in which all life forms shared a common origin.
4. In the mechanistic view, articulated by Descartes and many successors to the present day, life is no more than the result of very complex configurations of matter.
5. Analogously, many artificial intelligence researchers have thought that once computer chips achieve a density of interconnection comparable to the neurons in the human nervous system, consciousness will emerge.
6. In the vitalist view, life is special, marked by the action of forces that are not reducible to physics or chemistry and certainly not to mechanical action.

C. Lamarck was an early proponent of the unity of life but not the first or the only one.
1. George Buffon was an important predecessor a generation earlier.
2. Erasmus Darwin, Charles's grandfather, was Lamarck's contemporary.
3. Lamarck's evolutionary theory of life was a dynamic one that was highly influential in the 19th century, though reviled in the 20th.

II. The introduction of the microscope in the 17th century opened up the scientific study of life.

A. For one thing, the microscope revealed unseen and unseeable "worlds."
1. The researches of Anton Leeuwenhoek, Robert Hooke, and Marcello Malpighi expanded our conception of the realm of life forms.
2. The drive to see more by increasing magnification caused blurrier images and some researchers abandoned the use of the microscope as hopelessly misleading.

B. Correcting chromatic aberration allowed much greater magnification and much deeper glimpses of the secret of life.

1. John Dolland discovered how to combine different types of glass to correct aberration and took out a patent in 1758.
2. Contemporaneously, the great mathematician Leonhard Euler developed a mathematical theory for correcting aberration.
3. Euler's theory was not used to design commercial lenses for 100 years but then became the basis of the famous Carl Zeiss Company line of state-of-the-art microscopes.

C. Although the problem of chromatic aberration was solved by the late 1700s, it was not until the 1820s that a new generation of research microscopes became available.
 1. Giovanni Battista Amici developed a complete microscope "package" incorporating the new lenses, and this, along with rival designs, inaugurated a new phase in biological research.
 2. In effect, the cell theory of life "popped out" almost at once.
 3. There is an important lesson here about scientific instruments: The technical capabilities of their components are much less important than the usability of their integration into a functional design.

III. The microscope is inextricably involved with the cell theory of life and is a powerful case study in the relationship between instruments and theories.

A. The origins of the cell theory go back to the 17th century.
 1. Robert Hooke coined the term *cell* to describe the microscopic structure of slices of cork.
 2. In the 18th century, botanists speculatively generalized Hooke's observation to all plants.
 3. With color-corrected microscopes in the 1820s and 1830s, observers identified not only cells in plants and animals but even the nucleus within the cell (at that time, just a dark spot).

B. Cells were proclaimed the building blocks of all living things in the late 1830s.
 1. In 1838, Mathew Schleiden proclaimed the universal cellular structure of all plant tissue, and in 1839, Theodor Schwann did the same for animals.
 2. Both held mechanistic views of life but were associated with the fertile laboratory of Johannes Müller, who was a stout vitalist.

    3. For no logically compelling reason, the cell theory of life was effectively universally adopted as true immediately.
    4. By contrast, the atomic theory of matter was not accepted as true by many chemists for decades and by most physicists for 100 years after Dalton.
  C. Given the cell theory of life, the leading research subject must be to discover what is going on inside the cell.
    1. Fortunately, microscopes kept improving, especially from 1870 with the new Carl Zeiss models, as did specimen preparation techniques.
    2. One crucial question is: Where do cells come from?
    3. In the 1850s, Rudolf Virchow emerged as a champion of the cell as the building block of life and of the view that all cells come exclusively from other cells.
  D. This highlights the process of cell division going back to the fertilized egg as the "starter" cell of all sexually reproducing organisms.
    1. During 1850 to 1875, it was established that the fertilized egg is a single cell formed by the fusion of a sperm cell and an egg cell.
    2. In the 1880s, microscopes allowed observation of new details of the process of cell division, named *mitosis* by Walther Flemming, and the stages of this process were identified by the end of the century.
    3. Note two points for future reference in the lecture on gene theory: First, in the 1870s, Friedrich Miescher suggested that an acid molecule in the nucleus might play an important role in cell division. Second, the question of how organisms inherit characteristics from their parents now focused on the formation of the "starter" cell.
IV. The overarching question for biology, of course, is: What is life? The cell theory must be understood as a response to that question.
  A. That cells are the universal building blocks of life still does not tell us what life *is*.
    1. In the mid-19th century, it was proposed that the essence of life was *protoplasm*, a term coined by Hugo Mohl, collaborator of the very influential mechanist chemist Justus Liebig.

      2. By the end of the century, with the growing understanding of the complex structures and molecular processes internal to the cell, the colloid theory replaced the protoplasm theory.
      3. Early in the 20th century, enzymes replaced colloids as the key to life, that which controls the cellular processes that confer life on a mass of chemicals.
      4. With the discovery that enzymes are proteins, enzymes became the essence of life; this view endured into the 1950s.
B. The great German chemist Emil Fischer put enzyme theory on the research map.
      1. Fischer discovered that enzymes are composed of amino acids, which could readily be synthesized.
      2. He was able to synthesize proteins out of amino acid combinations, including proteins occurring in the human body.
      3. This strongly reinforced the mechanical/biochemical view of life, and vitalism disappeared as a scientifically respectable theory.
      4. In the 1950s, DNA was proclaimed the essence of life because it contained the "instruction set" for protein manufacture.
      5. DNA theory perfectly illustrates the atomistic style of thinking, at least in its initial development, but we'll return to this in the lectures on gene theory and molecular biology.

Essential Reading:

William Coleman, *Biology in the Nineteenth Century*.

Ernst Mayr, *The Growth of Biological Thought*.

Questions to Consider:

1. Is life ultimately "just" a matter of physics and chemistry?
2. Can the nature and behavior of organisms be explained at the cellular level?

Lecture Twenty
The Germ Theory of Disease

Scope: The germ theory of disease is the cornerstone of modern scientific medicine, yet it was quite controversial initially. From antiquity, the dominant view of illness was that it was the manifestation of an imbalance within the body, a view revived in 19th-century homeopathic medicine. The expertise of the physician lay in identifying the imbalance and restoring the balance. There were, to be sure, parallel traditions that some illnesses at least were divine afflictions or were caused by natural external agents, but the rational view was that the primary cause was an internal derangement. By the mid-19th century, a scientific view of illness arose that sought a materialistic-chemical explanation of illness, and the defenders of this view opposed the germ theory when it was proposed by Louis Pasteur, Robert Koch, and others. The triumph of the germ theory and the medical education, research, and treatment it provoked did not preclude materialistic and homeopathic causes of illness.

Outline

I. Why we become ill has been an issue from the beginning of recorded history.
  A. The oldest and most enduring answer is that illness is a divine visitation.
   1. Typically, illness is perceived as a punishment for moral or ritual misbehavior.
   2. This is reflected in the close connection between healing and the priesthood in ancient Babylon and Egypt but also in ancient Greece.
   3. Apollo was the Greek god of healing, and Apollonian temples/health clinics were built across the Greek world.
  B. Sick people came to these temples to pray and to receive treatment from the priest-healers, but naturalistic clinics arose, as well.
   1. In the 5th century B.C.E., Hippocrates established his medical school on Kos, while a rival school flourished in the city of Cnidus.

2. In the 2nd century C.E., Pergamum was the home of Galen, whose medical writings and theories remained influential for more than 1500 years.
C. Biblically, too, illness is described as a punishment and healing as coming from God.
1. In Exodus, God says, "I, the Lord, am your healer," and this theme recurs throughout the Hebrew Bible. Jesus is portrayed as a healer in the Christian Bible.
2. Illness as divine punishment was a commonplace throughout the Christian Middle Ages and persists in the present day.
D. The view that disease is a natural phenomenon, with natural causes to be dealt with naturally, was developed in the Classical period.
1. Hippocrates's *Airs, Waters and Places* discusses the effects of environment on health, while another book rejects the view that epilepsy is a divine affliction.
2. The Cnidian medical school was focused above all on the course of a disease over time, as if the disease were an entity in its own right, independent of the patient.
3. In the 1st century B.C.E., the Roman writer Varro, followed in the next century by the Roman writer Columella, wrote of some diseases as caused by "minute animals" entering the body.
4. In the 2nd century C.E., Galen's writings summarized and extended a tradition that health is a balance among the 4 "humours" of the body and illness an imbalance.
5. This tradition survived Galenic theories, reappearing in the 19th century as homeopathy, for example, and as the homeostasis theory of Claude Bernard.
6. The cause of disease in this still-influential view is a malfunction internal to the body.
E. The idea that the cause of disease is something external that invades and sickens the body preceded the germ theory.
1. Hippocrates and Varro and Columella were proponents of this idea in antiquity.
2. The response of quarantine to the outbreak of plague in 1347 suggests a conception of contagion spread from outside the body.

3. The Renaissance naturalist Girolamo Fracastoro suggested that syphilis, then epidemic, was caused and spread by invisible living things.
4. With the 17th-century microscope discoveries of invisible "animalcules" in water and bodily fluids, Leeuwenhoek wrote to the Royal Society suggesting a relationship between these creatures and illness.

II. In the 18th century, mechanistic theories of life took strong root, but it was in the 19th century that modern theories of disease emerged and, with them, modern medicine.
   A. In fact, what emerged was a battle between rival scientific theories of disease.
      1. On one side were the men commonly identified as the founders of the germ theory of disease: Louis Pasteur and Robert Koch.
      2. Their argument was that disease was caused by microscopic life forms entering the body and disrupting its normal processes.
      3. On the other side were their opponents, including some of the most prominent names in 19th-century science, among them, Justus Liebig, a defender of the mechanical view of life, and Rudolf von Virchow, for whom all abnormal functioning was cellular malfunctioning.
      4. The germ theory of disease is popularly depicted as one of the great triumphs of modern science, yet from our perspective today, the opposing view was right—or certainly not wrong!
   B. Pasteur anticipated the modern germ theory of disease in his famous paper of 1857 on fermentation.
      1. The idea that fungi and molds were causes of plant diseases and skin diseases in humans was argued in the decades before Pasteur's paper.
      2. Schwann had identified yeast as a living organism responsible for the fermentation of sugar solutions in the mid-1830s.
      3. Pasteur formulated an experiment-based theory of invisible organisms busily at work, constructively and destructively, in organic processes of all kinds, as effects of their life cycles and perhaps playing a role in disease, too.

4. As microscopes improved, bacteria became visible; viruses were identified in 1898 but were not visible until the electron microscope was introduced in the 1930s.
5. Opposition to the germ theory was led by Virchow using a journal he founded and a text he published to champion the cell as an "independent unit of life."

C. Jacob Henle's *Handbook of Rational Pathology* (1846–53) speculated that contagious diseases were caused and spread by microscopic parasite-like living creatures.
1. Henle had studied under Müller, was a close friend of Schwann, and was the teacher of Robert Koch.
2. He laid down three conditions for establishing scientifically that an invisible agent was the cause of a disease, and these became the hallmark of Koch's experimental method.
3. By 1876, Koch had isolated the anthrax bacillus and studied its complete life cycle; in 1882, he isolated the tubercle bacillus and, in 1883, the cholera bacillus.
4. Meanwhile, working in parallel, Pasteur had been studying anthrax, convinced now that microbes cause disease.
5. In parallel, Pasteur and Koch showed experimentally that this was indeed the case.

D. Pasteur resurrected vaccination and foresaw the commercial uses of microorganisms in industrial processes, in agriculture to control other organisms, and as antibiotics.
1. Pasteur's studies of cholera in 1879 "accidentally" revealed the efficacy of vaccination, initially against animal diseases, then in humans. [Note: Also in 1879, Pasteur isolated the streptococcus bacterium and claimed that it was responsible for puerperal fever; he also isolated staphylococcus, linking it to osteomyelitis.]
2. His 1881 public anthrax challenge led to the mass vaccination of sheep and cattle, and his rabies research isolated a virus.
3. In 1885, Pasteur suggested that dead bacteria could also confer immunity, and in 1890, two students of Koch showed that serum could confer immunity, specifically for diphtheria.

III. Opposition to these new theories of fermentation and disease was not at all irrational; in fact, it was correct.

A. Obviously, the germ theory of disease and the therapies it stimulated led to major improvements in public health.
  1. In the developed world, memories today are very dim—or nonexistent—of the awesome toll taken by smallpox, measles, typhus, cholera, diphtheria, yellow fever, and polio.
  2. Inspired by Pasteur, Joseph Lister, among others, promoted antisepsis to reduce the rate of infection caused by the hands of physicians and their instruments.

B. But there were scientific grounds for opposing the germ theory.
  1. Some people had identifiable germs in their blood but displayed no signs of illness.
  2. Others displayed signs of illness but without visible germs in their blood or tissues.
  3. The atomistic style of thinking here reveals its character: Either germs are *the* cause of disease or they are not.

Essential Reading:

Ernst Mayr, *The Growth of Biological Thought*.

Michael Worboys, *Spreading Germs*.

Questions to Consider:

1. Common sense strongly suggests that factors external to the body cause disease; why, then, was there opposition to the germ theory?
2. Are scientific theories ever the product of a "heroic" individual genius, or does the thinking of a collective find expression through individuals?

Lecture Twenty-One
The Gene Theory of Inheritance

Scope: Conceptions about inheritance are among the oldest ideas in recorded history and, already in ancient Greece, were connected to ideas about embryological development. In the late 18th century, inheritance was identified as central to a theory of the historical existence of, and contemporary relations among, plants and animals. It was central to the question of the fixity of species, for example; thus, every theory of evolution required a theory of inheritance. Gregor Mendel's idea that inheritance was determined by the combination of discrete agents—"atoms" of inheritance, each with a fixed property or influence—was therefore timely and has proven to be very powerful; yet it was ignored for decades. The recovery of this idea at the turn of the 20th century; its convergence with, and resuscitation of, evolution by natural selection; and its rise to dominance with the "decoding" of the DNA molecule, constitutes one of the great epics of science.

Outline

I. The gene became "real" for most biologists in the years after 1910, but not for all and not even today.

  A. There are interesting connections among the atomic theory of matter, the cell and germ theories, and the gene.

  1. The gene was conceived as a discrete unit of inheritance, a "black box" analogous to atoms, cells, and germs, each gene determining one phenotypic character.

  2. Wilhelm Johannsen, who coined the term *gene* in 1909, did not himself believe that the gene was an "atom" of inheritance.

  3. What Johannsen meant by *gene* was a kind of accounting or calculating unit useful as a name for the complex *process* of character transmission.

  4. Inheritance involved a distinctive encapsulated process within the cell, but it was not a matter of transmission of independent black boxes from parents that combine to produce the characteristics of their offspring.

5. This process, Johannsen believed, involved substances in the nucleus of the cell that interacted during cell division, but it was more complex molecularly than a fusion of black boxes.
- **B.** Genes first became real at Columbia University in 1909.
   1. Thomas Hunt Morgan was an opponent of genes until 1909, when he decided to study mutations in *Drosophila* and very soon discovered that an eye-color mutation could be localized on one chromosome.
   2. Localization implied physical reality for Morgan—as it would for the chemical and physical atom—and over the next six years, he and his graduate students identified more than 100 mutant genes and established the primary features of a gene theory of inheritance.
   3. The climax of this work came in the mid-1920s, with the success of Morgan's student Hermann Muller in artificially inducing mutations using X-rays.
   4. Muller's experiment, for which he received a Nobel Prize, was a "smoking gun" for the reality of genes because the prevailing quantum theory explained X-rays as photon "bullets": in order to cause mutations, the X-ray photon would have to strike a specific, identifiable site where the mutation would take place, namely, a localized, "atomic" gene!
   5. From the perspective of the present, given our growing understanding of how DNA fits into a network of cellular processes, Johannsen seems closer to the mark than Morgan.

II. As with disease, people have always been interested in the nature and causes of inheritance.
- **A.** The problem of inheritance is linked to the problem of explaining embryological development.
   1. One problem is explaining how it is that embryos develop so similarly to their parents, a constancy-of-type problem: Why is it that individuals reproduce "after their kind," as Genesis puts it?
   2. A second problem is individual variation: Why do offspring differ from their parents, typically in very small ways, but often significantly?
   3. Explaining embryological development would clarify the nature of species and genera, whether they are really features

of nature or conventional categories (recall the problem of universals).
- **B.** Ideas about inheritance have a long history, but they became scientific only in the second half of the 19th century.
  1. Hippocrates proposed a "pangenesis" account of inheritance not much less sophisticated than the one Darwin held.
  2. In this view, every characteristic of the offspring was represented in the sperm or egg: arms, legs, hair, and so on.
  3. This is reminiscent of Anaxagoras's version of atomism, in which there is a kind of atom for every kind of thing in the world, and organisms grow by separating out the atoms they need.
  4. Four hundred years later, Lucretius popularized this pangenesis theory in his poem *On the Nature of Things*.
  5. Aristotle, by contrast, argued that sperm was a formative cause inside the fertilized egg, hence responsible for the emergence of all the forms in the offspring, while the egg was the "stuff" that was formed by the influence contained in the sperm into a distinctive embryo.
- **C.** Inheritance theory, especially from the 18th century onward, developed in close association with embryology, taxonomy, and evolutionary ideas.
  1. The *preformationist* theory of embryological development explained that development away by denying that there was development: The embryo was fully formed in miniature in the egg or in the sperm.
  2. A rival view, *epigenesis*, argued that development was real, with new forms emerging out of the formless fertilized egg under some unknown forces, vitalistic or mechanical.
- **D.** The central problem of taxonomy, made urgent by the sudden discovery of thousands of new life forms, is whether classification is natural or conventional.
  1. If classification is conventional, then a "species" is just a convenient way of organizing a group of plants or animals that seem to be very similar to one another and that breed true to type.
  2. On the other hand, if classification is natural, then a species is a thing that really exists, and breeding true to type is a sign

that there is a natural system for classifying plants and animals.
3. Carl Linnaeus emerged in the late 18th century as the arch-defender of the natural position, having devised a classification system based on the sexual organs of plants.
4. Plant sexuality was a (scientific) discovery of the 18th century that remained controversial well into the 19th.
5. Linnaeus's classification system is still the standard for biologists, but ultimately he had to acknowledge that it was conventional.
6. The taxonomy problem lies at the heart of the central claim of evolutionary theory that species are not natural categories and that life forms continuously evolve.

III. The gene idea appears as a solution to the problems of inheritance, embryological development, evolution, and taxonomy.
   A. Enhanced-microscope studies of structures internal to the nucleus of the cell in the last decades of the 19th century led to modern genetic theory.
      1. Around 1880, Walther Flemming identified chromosomes within the nucleus as pivotal to cell division and the transmission of characteristics to the next generation of cells.
      2. Concurrently, August Weismann developed his *germ plasm* theory of inheritance keyed to a substance on the chromosome inherited from the parents.
   B. Modern gene theory is invariably called Mendelian, but Gregor Mendel had nothing to do with its emergence or development!
      1. Mendel's plant-breeding experiments, inspired by the pre-Darwinian evolutionary ideas of his biology professor at the University of Vienna, were published in 1865, but to no effect.
      2. In the late 1880s and again in the late 1890s, at least three botanists already committed to a discrete unit within the nucleus of the cell as the solution—an "atomistic" solution—to the problem of inheritance independently discovered Mendel's work and graciously credited him as their predecessor.
      3. These men were Hugo de Vries, the first and most important; Carl Correns; and Erich Tschermak. I think William Bateson,

who was last to discover Mendel's work, in 1900, also belongs in this group.
4. De Vries published his "Mendelian" theory of inheritance in 1889, three years before he learned of Mendel's work.
5. He subsequently developed a mutation theory to explain individual variation and evolutionary speciation (in place of natural selection).
6. It is De Vries, not Mendel, who founded modern genetic theory, because that theory surely would have developed as it did had Mendel's prior work never been discovered.
7. Mendel receives more credit than he deserves, just as Copernicus, too, receives credit for an idea that was developed into its scientific form by Kepler.

C. Gene theory evolved from a "something" in the nucleus to DNA in a little more than half a century.
1. At the end of the 19th century, this something was a speculative black box on a chromosome, then a real black box after Morgan's work.
2. From the 1930s, the black box was believed to be made up of enzymes, but in 1953, it was revealed to be the DNA molecule.
3. By 1960, the solution to the problem of inheritance and the allied problems of embryological development, taxonomy, evolution, and life itself was the "code" programmed into the DNA molecule.

Essential Reading:

Michel Morange, *The Misunderstood Gene*.

James Watson, *DNA*.

Questions to Consider:

1. With the fate of Mendel's work in mind, are scientific theories dependent on specific individuals or are they a product of a climate of opinion?
2. Are genes "the" answer to inheritance, or are they, in the end, important but not determinative of the complex process we call inheritance?

Lecture Twenty-Two
Energy Challenges Matter

Scope: Motion and forces have been central to the explanation of change—the root problem for natural philosophy since the ancient Greeks—but the ultimate reality until the 19th century was matter. Even for early modern science, matter was the subject of change, matter and its motions alone were conserved, and forces inhered only in matter. The metaphysical dominance of matter was undermined in the mid-19th century with the introduction of the idea of energy. Energy achieved formal recognition as a feature of reality parallel to matter in the new science of thermodynamics. Now energy, too, was conserved, and it was endlessly convertible among numerous immaterial forms. The idea of energy stimulated process theories in science in which patterns and relationships were real. By 1900, it was proposed that energy alone was ultimately real.

Outline

I. The fundamental features of reality for early modern science were matter, motion, and forces, but in the mid-19th century, energy was added to the list.

   A. The idea of energy as an elementary feature of reality required a conceptual reorientation for scientists.
   1. Energy does not fit well with the atomistic style of thinking.
   2. The 19th century witnessed the flowering of this atomistic style in the atomic theory of matter, the cell theory of life, the germ theory of disease, and the gene theory of inheritance.
   3. The creation of a new science, thermodynamics, in response to the idea of energy posed a challenge to mechanics, the reigning science of matter in motion.
   4. Thermodynamics forced a reconceptualization of time, because the role of time in thermodynamics is inconsistent with the role that time plays in mechanics.
   5. There is a peculiar irony here—what philosophers might call a Hegelian irony—in the atomistic style of thinking being undermined at the height of its apparently decisive triumph.

- **B.** Recall that the core problem for science is explaining change.
  1. The Parmenidean/atomistic approach to this problem is to reduce change to the interactions of things with changeless properties.
  2. The Heraclitean/process approach to explaining change is to accept change as real and irreducible, and to seek the patterns or laws of change.
  3. For the process approach, the goal of the scientist is to identify the distinctive properties, not of elementary things, but of processes.
- **C.** The focus of modern science, from Descartes on, was on explaining all natural phenomena in terms of matter in motion.
  1. From the time of Descartes to the mid-19th century, the center of attention was matter and the properties it possessed, including intrinsic forces, that explained its motions.
  2. In the mid-19th century, highlighted by the idea of energy, the center of attention became the motions of matter and the patterns or laws of those motions.

II. Heat may seem too prosaic a phenomenon to have provoked such a profound reorientation, but it did.
- **A.** Since the 17th century, attempts to determine what heat *was* proved elusive, in spite of its apparent simplicity.
  1. One view was that heat was motion and, thus, no *thing* at all.
  2. The rival view was that heat was the escape of a weightless, invisible thing called *caloric* from an object.
  3. No convincing case was made by one side or the other through the 18th century.
- **B.** By the 19th century, disputes over the nature of heat had become bitter.
  1. A new urgency to understand the nature of heat was lent by the commercialization of the steam engine beginning in 1775.
  2. An attempt to resolve the dispute experimentally by Benjamin Thompson failed, in spite of its description as a success in science texts to the present day.
- **C.** Thompson was convinced by his cannon-boring experiment, but many eminent figures were not.
  1. Pierre-Simon Laplace held the caloric view to the end of his life.

2. John Dalton, whose book founded the modern theory of the atom, wrote about as much in that book about caloric as about his atomic theory of matter.
2. Furthermore, Sadi Carnot, an engineer who produced the first scientific analysis of heat engines, also supported the caloric theory.

III. What Carnot contributed to a scientific revolution was every bit as profound for science as the industrial revolution was for society.
   A. Carnot set out to measure the efficiency of a heat engine.
      1. Using the caloric theory of heat as his framework, he sought and found a quantitative relationship between the fuel burned in a steam engine and the amount of mechanical work produced.
      2. Carnot found a limit to the efficiency of any heat engine, regardless of the design, that is still accepted today.
      3. In the course of this analysis, Carnot had to define *work* as a *scientific* idea.
      4. Fortunately for Carnot, Émile Clapeyron formulated his mostly qualitative ideas mathematically, bringing them to the attention of physicists in 1830.
   B. Carnot's ideas belong in the broader context of a discussion of forces in early-19th-century physics.
      1. Descartes' mechanical philosophy of nature allowed only contact forces among moving material objects.
      2. All natural phenomena had to be reduced to the consequences of collisions among moving matter.
      3. Newton embraced a range of non-contact forces, forces acting at a distance, to explain gravity, electricity, magnetism, optical phenomena, and chemical combination.
      4. Non-contact forces seemed the only way to explain the full spectrum of natural phenomena.
   C. The growing list of forces that, by 1800, were scientifically respectable inevitably raised the question of relationships among them.
      1. Were there relationships among heat, light, electricity, magnetism, gravity, mechanical forces, and selective chemical combination?

2. The electric battery proved an important new scientific instrument, replacing spark-generating devices.
   3. Humphrey Davy used continuous current to decompose molecules into their elements, revealing a relationship between electrical and chemical forces.
   4. Sadi Carnot showed a relationship between heat and mechanical work.
   5. Hans Christian Oersted showed that electricity and magnetism were connected.
   6. Oersted subscribed to the so-called Romantic nature philosophical movement, whose members believed in the fundamental unity of nature and, thus, the ultimate unity of all natural forces.
   7. Michael Faraday extended Davy's work in electrochemistry and Oersted's observation of a connection between electricity and magnetism.
   8. Faraday invented the dynamo, which connects mechanical action (motion), electricity, and magnetism, and he showed that both electricity and magnetism affect light, suggesting that light is an electromagnetic phenomenon.

IV. These developments converged in the 1840s to produce the new science of thermodynamics.
   A. The occasion for the convergence was the study of heat.
      1. In the 1840s, James Prescott Joule did a series of carefully controlled, quite simple experiments to determine precisely Carnot's mechanical equivalent of heat.
      2. Like Carnot, Joule supported the caloric theory of heat, but unlike Carnot, he did not believe that caloric was conserved.
      3. William Thomson, later Lord Kelvin, was aware of Carnot's work (through Clapeyron) and of Joule's work, but he was inclined to the view that heat was motion.
      4. Motion is conserved, so if there is a precise relationship between heat and motion, then *something* in this process is conserved.
   B. While Joule and Thomson were studying heat, so were a number of Continental scientists, who also were looking for correlations among the forces of nature.

1. In 1847, Hermann Helmholtz published an essay claiming a new conservation law: the conservation of *Kraft*.
2. *Kraft* then meant "force" or "power," but in 1851, William Rankine co-opted the word *energy* to describe what was conserved and gave it a scientific meaning.
3. Thomson called the concept of energy the most important development in science since Newton's *Principia*.
4. From 1850, Rudolf Clausius formed the new science of thermodynamics around an expanded conception of heat as a form of energy.
5. The first law of this new science was that energy was conserved, and the second was that energy could flow only from a hotter body to a colder one, never the reverse.

C. Clausius identified a mathematical quantity he called *entropy* that was a measure of this irreversible flow of heat.
  1. In 1853, Thomson announced that this idea implied that time was directional and irreversible, contrary to the role of time in mechanics.
  2. Clausius proclaimed as a law of nature that entropy must increase in any closed system; this was soon taken to imply the "heat death" of the Universe.
  3. This is reminiscent of the Newton-Leibniz controversy over whether the Universe needs to be "wound up" by God every now and then to prevent it from just this kind of death.

D. The idea of energy is a fundamentally new *kind* of idea in modern science.
  1. Energy is real and energy is conserved, but there is no such *thing* as energy!
  2. Energy occurs only in specific forms, but those specific forms are not conserved: They are interconvertible.
  3. Note that we have no experience of energy in and of itself, nor can we have such an experience.
  4. This enhances the distinction between atomistic science and process science, of which energy is a prime example.
  5. By the late 19th century, it was proposed that energy, not matter, was the ultimate reality.
  6. Furthermore, the energy idea was only one instance of process thinking in 19th-century science.

Essential Reading:

Y. Elkana, *The Discovery of the Conservation of Energy*.

P. M. Harman, *Energy, Force and Matter*.

Questions to Consider:

1. If *matter* is a name for certain stable energy patterns, what is the "stuff" of which energy is made?
2. Is the substantiality of the Universe an illusion, analogous to the illusion of continuous motion in motion pictures?

Lecture Twenty-Three
Fields—The Immaterial Becomes Real

Scope: It is ironic that the apparently climactic development of modern science in the 19th century saw the foundation of its conception of reality—materialistic determinism—undermined by developments internal to science. Energy was one immaterial reality, and electric, magnetic, and electromagnetic fields were another, soon to be supplemented by the aether field and field theories of gravity. The idea of the field went deeper into the metaphysics of modern science than adding new immaterial realities, however. The seemingly insurmountable difficulty of formulating a plausible physical mechanism for the action of fields, for how they transmit energy and forces, led to a reconceptualization of the nature of scientific theories. It became increasingly orthodox to argue that scientific theories described human experience, that their truth was not a function of correspondence with a reality existing independent of experience, but of our evolving experience of how nature behaves.

Outline

I. The introduction of fields as elements of physical reality in the 19th century was epochal for the evolution of modern science.

A. The idea of the field is related to the idea of energy.
1. Energy is real, but it is not material, hence not a thing in the traditional sense.
2. If it is not material, it should not be real, according to the dominant modern scientific view that all natural phenomena are caused by matter in motion, as in Laplace's famous declaration of the exhaustiveness of materialistic determinism.
3. Energy is immaterial but lawful action in the spirit of Heraclitus: to understand nature is to understand its *logoi*, its laws or rules.

B. But the law that energy must be conserved, that all forms of energy are interconvertible only in rule-governed ways, does not tell us *how* energy acts.

1. How is electrical energy, for example, transmitted within a conductor or propagated through space?
2. The broader issue is the continuing concern over the nature and the reality of non-contact forces, forces that act at a distance but are nevertheless consistent with science, as Newton said, not magical, as the Cartesians said.

II. The central role in the idea of fields of energy and force as the solution to how energy and forces act was played by a self-educated, working-class Englishman named Michael Faraday.

A. Faraday's researches in electricity and magnetism, largely qualitative because he was weak in mathematics, brought the issues latent in Newtonian action-at-a-distance physics to a head.
1. Electromagnetic induction led Faraday to theorize about how electric and magnetic forces propagated through space and through conducting bodies.
2. Rejecting both the reality of a vacuum and action at a distance, Faraday speculated that "electrical bodies" were surrounded by a "dielectric" medium that, when stimulated by a flow of current, created a state of "electric tension" transmitted through the medium that caused a current to flow in a nearby body.
3. In fact, he proposed that all of space was filled by a form of matter for which he adopted an ancient Greek name, *aether*.
4. The aether (or ether) constituted an absolute backdrop for ordinary matter and its interactions, its contiguous particles carrying and propagating electric, magnetic, and electromagnetic forces and light.

B. From his electrochemical studies, Faraday concluded that molecules were electrically polarized and, thus, responsive to electrical forces.
1. For years, Faraday gave different accounts of electrical and magnetic phenomena: mathematical, in terms of lines of force, and physical, in terms of the action of polarized contiguous particles.
2. Faraday believed that the lines of force were physically real and sought, unsuccessfully, a theory that described them as embodied in matter.

  3. In 1844, Faraday definitively gave up the atomic theory as physically real.
  4. His position was that matter was continuously distributed in, but not identical with, space, which was property-less.
 C. In rejecting atomism, Faraday explicitly denied the impenetrability of matter and its indivisibility.
  1. In 1845, Faraday used the term magnetic *field*, which quickly caught on, and in 1846, he proposed a physical interpretation of his lines of force as reflecting the structure of matter.
  2. Faraday was influenced by experiments showing that magnets could rotate a light beam, suggesting that light was, in fact, an electrical phenomenon and that the lines of magnetic force were physically real.
  3. This conclusion was further reinforced by his 1852 experiments with iron filings, which he considered a definitive demonstration of the physical reality of magnetic energy fields and the means by which they acted.

III. Faraday's ideas were developed by others with better mathematical skills, especially William Thomson and James Clerk Maxwell.
 A. Thomson was a major influence in science generally and in public affairs affecting science.
  1. In the 1840s, Thomson noted the implications of Fourier's "analytical" theory of heat and an analogy between the form of Fourier's equations for the flow of heat and the equations describing Faraday's electrostatic forces.
  2. This suggested the possibility of a physical analogy, as well; that is, he asked: Do electric and magnetic forces "flow" through a medium analogous to the flow of heat by way of contiguous material particles?
  3. Fourier's example notwithstanding, Thomson devoted decades to the search for a mechanical model of the aether from which its functions as carrier and transmitter of non-contact forces could be deduced.
 B. The idea of the field reached maturity in the work of James Clerk Maxwell.
  1. Maxwell took up the problem of a physical model of the aether that would satisfy the criteria of Faraday's field concept.

2. The problem was complicated by the need for the aether to carry and propagate non-contact forces without absorbing any of their energy or interfering with the motion of the planets.
3. In addition, a new wave theory of light had displaced Newton's corpuscular theory by the 1850s, and the aether was a "natural" medium for those waves, but it now also had to transmit light waves without affecting them in any way.

C. In 1865, Maxwell published a truly epochal work in physics titled "A Dynamical Theory of the Electromagnetic Field."
1. Maxwell united electricity, magnetism, and light into a single mathematical framework, interrelating them by a single set of equations from which all relevant phenomena could be deduced.
2. The field was now firmly established as physically real, but Maxwell had "settled" for a strictly mathematical description of it, à la Fourier's theory of heat.
3. In 1873, Maxwell argued that there were an "infinite number" of hidden physical mechanisms that could generate observed behaviors, while a single mathematical description could be confirmed empirically.
4. The mathematical theory satisfies all the empirical requirements without specifying the physical mechanism involved.
5. Note well, however, that regardless of the physical mechanism, the field is fundamentally incompatible with atomism.
6. Where the atom is localized in space and discrete, the field fills all of space and is continuous.

D. Maxwell could not altogether give up the underlying physical mechanism.
1. Descartes had argued that knowledge of nature had to begin with hypothetically assumed premises because God could have created Nature in an infinite number of ways.
2. Maxwell noted that an infinite number of physical mechanisms could generate the relevant observed phenomena, but after all, the aether-field *is* real and must have *some* form, so he continued to attempt physical models!

- **E.** In the last third of the 19th century, there was near unanimity among physicists that the aether was the physical realization of real fields that transmitted energy and (non-contact) forces.
    1. There was vigorous disagreement among physicists over the nature of the aether, its physical form, and structure, but no disagreement over its reality right into the 20th century.
    2. The Irish physicist George FitzGerald championed an electromagnetic aether against the material theory of the aether proposed by Thomson (by then, Lord Kelvin).
    3. FitzGerald argued that the electromagnetic aether was the ultimate physical reality, with matter merely a stable electromagnetic energy pattern, or "knot," in the aether.
    4. This is, of course, a complete turnabout from the atomistic style of thinking because now atoms *become* energy!
- **F.** Physicists attempted to measure the physical effects of the aether, especially the Earth's velocity relative to it.
    1. Sophisticated experiments by Albert Michelson and Edward Morley using a new instrument of Michelson's design yielded zero.
    2. FitzGerald and Hendrik Lorentz, independently, explained this by noting that if both the aether and all matter are electromagnetic, then matter in motion and the aether must interact in such a way as to cancel the effects sought.
    3. This solution made the electrodynamics of moving bodies a central problem in physics, and in 1905, Einstein published a paper that confirmed the mathematics of the solution but dissolved the aether.

Essential Reading:

P. M. Harman, *Energy, Force, and Matter*.

Mary Hesse, *Forces and Fields*.

Questions to Consider:

1. Practically speaking, what difference does it make to us if scientific theories are true because they describe reality or because they describe experience and tell us nothing about reality?
2. Is it enough to say that a scientific theory is true because it "works" in practice, allowing us to control experience?

Lecture Twenty-Four
Relationships Become Physical

Scope: From Aristotle to the mid-19th century, relationships were considered predicates of subjects and had no reality in their own right. This is consistent with Parmenidean substance metaphysics and its expression in atomism in its various manifestations. The attribution of reality to energy and fields challenged substance metaphysics from the side of process metaphysics, but energy and fields seem to have a "tangible" character, even if immaterial. Not so relationships, yet by the end of the century, they, too, were recognized as causally efficacious and possessing properties and reality in their own right. Chemists discovered that the arrangement of atoms within molecules, including their spatial arrangement, determined the physical and chemical properties of molecules. The invention of symbolic logic provided a calculus of relationships and their inherent properties. Molecular biology, especially the decoding of the DNA molecule; network theory; and information theory showed that these relationships were physically real, further reinforcing process metaphysics.

Outline

I. What makes the idea of structure as a feature of physical reality important is that it conflicts with the historic preference for identifying reality with "thinghood."
   A. The reality of immaterial fields survived the dissolution of the aether into unreality.
      1. That fields are fundamental features of physical reality, on a par at least with energy and matter, is essential to 20th-century physics.
      2. Fields remain real even when their supposedly necessary substrate, the aether, itself conceived as either material or electromagnetic, drops out of reality.
      3. We have now seen both energy and fields become firmly established as immaterial realities within ostensibly materialistic 19th-century science.

    4. This challenges atomistic thinking by making it necessary to identify processes and their properties or laws.
    5. Energy and fields become, in effect, immaterial "substances" in an explanatory sense.
  B. It would be a mistake, however, to see the rise of process thinking as a debunking of atomistic thinking.
    1. The theories we have discussed that reflect what I have called a "Parmenidean approach" to explaining nature are very powerful theories: They unquestionably work.
    2. The theories that reflect what I have called a "Heraclitean approach" are also powerful and also work.
    3. A predilection for either/or thinking makes many feel that only one of these can be the "right" approach, when some combination of the two may be correct.
  C. Even more than the idea of energy or the idea of fields, the idea of relationships—relationships of *structure*—pushes the scientific attribution of reality to immaterial agents.
    1. Understand that structure here means relationships *among* the parts of an object, quite apart from characteristics inhering *in* each part.
    2. Energy and fields seem to have at least a quasi-thing-like character, but a relationship is *really* abstract.
    3. The growing recognition of process thinking—the recognition among mid-19th-century chemists that structural relationships have properties of their own, coming at the same time as the recognition of energy and fields—seems, to use a Freudian term, like a "return of the repressed."
II. The story of the idea of structure begins in France.
  A. In 1837, Auguste Laurent proposed what he called a *nucleus theory* of molecules.
    1. Laurent was a young chemist who had studied under the famous chemist Jean-Baptiste Dumas.
    2. Laurent, adopting the atomic theory of matter, proposed that certain molecules had properties that were a function of the geometric arrangement of their constituent atoms.
    3. At the time, chemists assumed that molecular properties derived from the constituent atoms and, for those who adopted

the atomic theory, from the precise proportions among those atoms: CO_2, H_2SO_4, and so on.

4. Laurent's idea was that molecules could have their atoms arranged in geometric patterns, such that the pattern defined a "family" of compounds with similar properties—*even if* the individual atoms were exchanged with different atoms, as long as the pattern was preserved.

B. This came to be called a *substitution theory*, and Laurent's professor, Dumas, adopted it and aggressively promoted it.
1. Dumas extended the theory by identifying many more compounds that possessed this property, and Laurent's nucleus theory was, to his chagrin, soon called Dumas' *type theory*.
2. Both were ridiculed for decades by some chemists, in particular by Wöhler and Liebig, who dismissed the idea that spatial arrangement could have physical properties of its own.
3. Many other chemists, ironically including some of Liebig's best students, incorporated the type theory into their research, achieving important results.
4. Note that the attribution of physical significance to spatial arrangement of atoms within a molecule implicitly strengthens the case for the reality of atoms, as opposed to treating them only as convenient heuristics.

C. The next step in making structure real came from Louis Pasteur, who worked with Laurent when they were graduate students and, like Laurent, studied under Dumas.
1. Pasteur's first research project was provoked by an 1844 report that some tartrate crystals—wine dregs—were optically active and some were not, even though both kinds were chemically identical.
2. Pasteur discovered in 1849 that *all* tartrate crystals are optically active but that the acid crystallizes into one of two forms that rotate light in opposite directions.
3. It follows that the same atoms have different spatial arrangements in these molecules and that the spatial arrangements cause different physical properties.
4. In the late 1850s, Pasteur noted that penicillin mold preferentially feeds off just one of these forms and speculated

that the three-dimensional arrangement of the atoms, perhaps as a helix, was important.
    5. The upshot of all this was that to explain the properties of molecules, you needed to know the atoms of which they were composed, the numbers of each atom, and how those atoms were arranged in space.
    6. That is, you needed to know the effects that were caused by a relationship.
  D. From the mid-1840s, the research programs of chemists following the Dumas-Laurent type/nucleus theories produced important results and created the field called *stereochemistry*.
    1. In the 1860s, August Kekule and his students showed that aromatic hydrocarbons constitute a family of compounds based on a single structure: a hexagonal "ring" of carbon atoms.
    2. The properties of each member of the family are determined by which atoms are attached to the carbon atoms and at which points.
    3. Stereo-chemical research revealed that carbon atoms can form long chains; this became the basis of polymer chemistry and the plastics industry.

III. The idea of structure/relationships as an elementary feature of nature is far deeper than chemistry, though it revolutionized chemistry and its applications.
  A. The abstract character of structure becomes clear in the case of symbolic logic.
    1. Logic was dominated for more than 2200 years by Aristotle's subject-predicate formulation, in which only subjects have properties and relationships are "accidental."
    2. As we have noted, though, deductive inference was strictly a function of form.
    3. The invention of a symbolic notation for reasoning had as powerful an effect on 19th-century studies in logic as the invention of a symbolic notation for mathematics in the 16th and 17th centuries.
    4. Augustus de Morgan, George Boole, Charles Sanders Peirce, Giuseppe Peano, and Gottlob Frege are among the pioneers of this innovation.

B. Almost immediately, symbolizing reasoning led to the recognition that relationships possessed properties of their own, independent of the terms (*relata*) they related!
   1. Given a relationship, we can begin to identify its logical properties regardless of who or what is related.
   2. Some of these properties are transitivity, symmetry, and reflexivity; we can determine the logical properties of the parent relationship, for example, without knowing who the parent and child are.
   3. George Boole's *Laws of Thought* was a particularly seminal contribution to symbolic logic.
   4. Boole's symbolic notation for how we think when we reason logically was soon recognized as capable of being implemented in electrical circuits.
   5. It took another 70 years before Claude Shannon did that, but when he did, he created the designs for the logic units built into all computer chips.

C. The study of relationships in and of themselves became characteristic of many major developments in 19th- and 20th-century mathematics.
   1. Differential geometry, a kind of highly generalized, abstract geometry, is an example, and it became important to quantum theory.
   2. The study of topology, that is, of freeform spatial relationships, anticipated by Leibniz in the late 17th century, is another example.

Essential Reading:

I. Grattan-Guinness, *The Norton History of the Mathematical Sciences*.

Mary Jo Nye, *Before Big Science*.

Questions to Consider:

1. How can we use language to refer to the world if meaning is strictly a function of relationships internal to the language?
2. How can scientific theories give us power over experience if theories are like languages, the meanings of their terms being a function of relationships internal to the theory?

Lecture Twenty-Five
Evolution as Process Science

Scope: Like Copernicus's moving-Earth theory before it and Einstein's theories of relativity after, the idea of evolution has not tangibly changed our daily lives and has led to no new technologies, but it has had a profound impact on our self-understanding. From the early 17th century to the mid-19th century, the scope of modern science expanded, naturalizing the Earth, the Solar System, the Universe as a whole, life processes within organisms, and with the advent of organic chemistry and the cell theory, life itself. The Darwin-Wallace idea of evolution by natural selection, reinforced by the contemporaneous discovery of ancestral human forms, enfolded humanity within the natural, a move completed by the naturalization of the nervous system and consciousness. Evolution has proved a powerful, innovative, and widely applicable cross-disciplinary idea, bringing contingency into scientific explanation, naturalizing apparent purposiveness, making time the dimension of novelty, and showing how novelty can emerge from the introduction of minute discontinuities into an underlying continuity.

Outline

I. The idea of evolution as formulated by Charles Darwin and Alfred Russel Wallace was a powerful and fertile intellectual innovation.

   A. Although famous as a theory of the "transmutation" of species, Darwin-Wallace evolution incorporates a transformed idea of time.

   1. We have already seen that the reigning idea of time—first incorporated into the deduction-based idea of knowledge from antiquity, and then into modern science—was that time *is not* a feature of reality.

   2. Time matters to us, of course, but not, in this view, to the absolute reality that is the ultimate source of experience.

   3. However, thermodynamics implied that time in physics was irreversible and a feature of reality, not merely an "illusion."

   4. Concurrently, Darwin and Wallace proposed that time in the biological world was irreversible.

        5. The evolution of species was the evolution *in time* of novel life forms.
    B. Evolution attributes explanatory power to a process, the process by which life forms are transformed.
        1. The process is *natural selection*, acting on continuous variability between parents and offspring under environmental pressure caused by overpopulation.
        2. There's a hierarchy of processes at work here, from the molecular processes underlying inheritance and cell division to survival of organisms in continually changing environmental challenges.
        3. Variability plays a key role, interacting with the struggle for survival and natural selection: Without variability, there is no evolution.
        4. Evolution is a complex process, in which the properties of the participants continually change over time, contrary to the fixed properties of literal and metaphorical atoms.

II. That species transform into new species was at the heart of a bitter controversy millennia before Darwin.
    A. Taxonomy may seem prosaic, but it is central both to the relationship between knowledge and reality and to the theory of evolution.
        1. The taxonomy problem is intimately related to the problem for which evolution is proposed as a solution.
        2. If species are constant because they are natural categories, then the problem of the origin of species takes one form.
        3. If species are conventional, then the problem takes a very different form.
        4. For example, the platypus, which combines mammalian features with non-mammalian features, challenges the category "mammal."
        5. The status of the names we use for grouping together individual objects reflects Plato's exhortation to "carve nature at its joints."
    B. With the European voyages of discovery, European naturalists were overwhelmed with thousands of new kinds of plants and animals.

1. Classifying a massive influx of individual objects amounts to an exercise in naming grouping features.
2. For some, the challenge is to name grouping features that are "real," that express the way nature is; for others, the challenge is to name features that are effective for botanists and zoologists.
3. The most compelling claim to a natural system was made by Carl Linnaeus, based on the sexual organs of plants, but in the end, he had to concede defeat.

III. If species names are conventional, then the constancy of species is an illusion.

   A. Darwin and Wallace were among those for whom species names were conventional.
   1. Species categories are really names for stable, but not permanent, groupings of continually (but slowly) changing varieties.
   2. Individual living things continually vary from their parents and from one another in one feature or another, while also sharing most of their features with their parents and some other living things.
   3. Over long periods of time, this variation may result in differences that lead to grouping individuals into different categories than their ancestors were grouped into.
   4. In the second half of the 19th century, biologists began to identify features of the processes of cell division and sexual reproduction that were causes of the variation at the molecular level.

   B. Artificial breeding is a familiar example of how new varieties are systematically created by people.
   1. For Darwin, but not for Wallace, this was analogous to the process underlying change in nature.
   2. For both, though, species names are names for varieties that are stable over what seem like long periods of time to human beings.
   3. Individual variability is, thus, the cornerstone of the idea of evolution, but note that the individual does not evolve!

   C. The question of what does evolve, if not the individual, leads to a deeper appreciation of the nature of evolution as a process theory.

1. Evolution by natural selection became dominant after the 1920s with the rise of population genetics.
2. Using newly developed statistics-based mathematical models, geneticists could calculate how gene changes would spread in a population over time as a function of the survival advantage or disadvantage they conferred.

D. Populations change their character over time as the distribution of genes changes, until the population looks very different from the way it had looked.
1. Combining Mendelian genetics, mutation theory, natural selection, and since 1953, a growing understanding of how DNA functions in the cell leads to a "picture" of how life forms radiate and adapt in time.
2. This is a "motion picture," but the rate of its motion is such that to us, each frame looks like a single photograph projected onto the screen and held there.
3. Run backwards, we would see a common origin for all life forms, but Darwin was initially cautious about this.
4. In the 18th century, Lamarck and Charles's grandfather Erasmus had argued for a single original life form and for processes that produced the incredible multifariousness of life forms that we observe.
5. Lamarck's theory involved a process of inheritance affected by the use/disuse of individual organs in the course of struggling to survive.
6. The origin-of-life question fits into the quest of the Romantic philosophy of nature to discover the ultimate unity of all nature, whether life forms or physical forces.

IV. In context, then, an evolutionary theory of life seems "natural."
A. If only from Darwin's critics, we know that the Darwin-Wallace theory of evolution did not suddenly pop up out of nowhere.
1. William Wells published an essay in 1813 suggesting a theory that we would recognize as evolution by natural selection.
2. This essay was described at length in a book by John Herschel that we know Darwin owned and had read.
3. In 1831, Patrick Mathew published a book on Royal Navy uses of trees that contained a theory of evolution based on the struggle for survival and selection.

4. In the 1840s, Robert Chambers published (initially anonymously) a British bestseller, *The Vestiges of the Natural History of Creation*, that presented a neo-Lamarckian theory of evolution.
B. The Darwin-Wallace theory became public in 1858.
 1. Darwin had been working out his theory since 1835, filling notebook after notebook with data, ideas, and arguments for species as types that change over time.
 2. Fortunately for him, he shared these ideas and access to some of his notebooks with leading British biologists.
 3. In 1858, Wallace sent Darwin the draft of a short essay, asking Darwin's opinion of his theory, based on biogeography, of the origin of species in terms almost identical to Darwin's.
 4. Darwin quickly wrote up a short paper of his own, and both papers were presented in 1858, followed by publication of *The Origin of Species* in 1859.
C. Evolution is a process theory, but it has two other important consequences.
 1. Evolution entails making time a fundamental feature of reality and irreversible, too, moving in one direction only.
 2. Individual variation is pivotal to evolution, and Darwin referred to the causative process as "spontaneous" but not "chance," meaning not random.
 3. Darwin's physicist contemporaries, however, saw his theory as implying that individual variation *was* a chance/random phenomenon.
 4. Evolutionary theory thus promoted yet another new scientific idea: that chance is a fundamental feature of nature, and thus, probability and lawfulness/rationality must be compatible.

Essential Reading:

Charles Darwin, *On the Origin of Species*.

Loren Eiseley, *Darwin's Century*.

Questions to Consider:

1. If human beings are wholly natural in all respects, how can human existence be more meaningful than the existence of any other organism?
2. Is it life that evolves or the whole system of which life is a part, and what are the boundaries of that system: the Earth, the Solar System, the Milky Way, the Universe?

Lecture Twenty-Six
Statistical Laws Challenge Determinism

Scope: Modern science was based on a deterministic conception of nature and natural explanation. With deduction the paradigm of reasoning about natural phenomena, it followed that every natural phenomenon is a necessary effect of a unique cause operating in accordance with universal laws. This was proclaimed by Laplace at the turn of the 19th century as a foundational principle of science. But in the course of that century, this principle was undermined by the introduction of statistics into scientific explanation and by the claim that certain processes, from the behavior of atoms to the behavior of human beings, obeyed statistical laws. The discovery of radioactivity led to the idea that radioactive "decay" was a stochastic process, and this was reinforced by mutation theory in genetics and the development of quantum mechanics. Earlier, however, a compelling case for statistical laws was made by social scientists and by physicists formulating the kinetic theory of gases and statistical mechanics and thermodynamics.

Outline

I. Over the last three lectures, we have seen the rise of a process style of thinking about nature challenge the dominant atomistic style.

   A. Concurrently, there was a challenge to a much deeper principle of modern science than atomism, namely, determinism.
      1. As we have seen, determinism is a corollary of the deduction-based conception of knowledge on which modern science rests.
      2. Effects flow from causes in nature as conclusions flow from premises in deductive arguments: necessarily.
      3. Determinism, whether strictly materialistic or incorporating energy, fields, and non-contact forces, is inescapable.

   B. To claim that natural phenomena have an intrinsically stochastic character has far-reaching implications.
      1. If nature is stochastic, then the same must be true of theories that claim to describe nature.

2. It follows that knowledge of nature cannot be universal, necessary, and certain, and neither can (all) laws of nature.
3. In fact, if nature is stochastic, we need to redefine what we mean by the terms *knowledge*, *truth*, *law*, and *reality*.
4. Yet Galileo's *Dialogue* reveals the depth of the commitment to knowledge as deductive and, hence, universal, necessary, and certain.

C. Probability theory emerged first in mathematics in the 16th century.
1. Cardan concerned himself with gambling probabilities, and in the 17th century, others developed mathematical theories of games of chance.
2. A turning point was Jacob Bernoulli's *Art of Conjecturing*, which claimed that probability theory could be applied to real-world decision-making.
3. That is, acting on the basis of probabilistic reasoning could be rational, alongside deductive reasoning.

D. Laplace was famous for defending the determinism of science, but he was also a contributor to probability theory.
1. Laplace saw probabilities as a reflection of human ignorance.
2. Faced with a situation *that has strictly deterministic causes* but about which we lack information, probability theory allows us to estimate the relative likelihood of outcomes given the information that we do possess.
3. This does not imply that *nature* is stochastic.

II. In the course of the 19th century, scientists had to consider the possibility that there was an irreducibly random character to at least some natural phenomena.

A. The Darwin-Wallace theory of evolution was one theory forcing this consideration.
1. Random mutations are an essential part of evolutionary theory today, and numerous physicists and philosophers thought the theory required that individual variation be random.
2. This is revolutionary in intellectual terms, but neither physics nor biology was leading the way.
3. In 1835, Adolphe Quetelet published a treatise on human behavior that introduced the then-radical idea of *statistical* laws.

B. Quetelet came to social science from the "hard" science of astronomy.
   1. He began his career in astronomy and learned the mathematical techniques of observational error correction that astronomers used, including the so-called "normal" distribution of values.
   2. His innovation was to apply these techniques to the growing body of statistical data about people that governments were collecting, a legacy of Bernoulli's idea that probability theory could lead to rational social decision-making.
   3. Quetelet saw that there were regular patterns of behavior among large numbers of people, for example, the number of suicides per year, or even per season or week, in a country or large city.
   4. These patterns, he claimed, allowed accurate predictions about groups of people quite independently of which individuals would do what or why.

C. This led Quetelet and, after him, Henry Buckle in England to proclaim the existence of statistical laws.
   1. On the one hand, this idea, on the face of it self-contradictory, reinforced the collection of social statistics by governments and by individual researchers.
   2. On the other hand, the idea of statistical laws made a deep impression on several physicists, among them, James Clerk Maxwell, who thought this idea could rescue human free will from the determinism of natural science.

D. At this very time, Maxwell and others were developing statistical theories in physics.
   1. Maxwell and Ludwig Boltzmann played lead roles in formulating the kinetic theory of gases, which describes the behavior of gases as a statistical process involving the behaviors of vast numbers of atoms/molecules.
   2. Both were also involved in a statistical interpretation of the second law of thermodynamics, which implies the irreversibility of time.
   3. On their interpretation, this is only a probabilistic law, not a necessary one; thus, the reversibility of time in mechanics is preserved, at least in principle.

4. The mathematics they developed subsequently was applied to mechanics itself, leading to the field of statistical mechanics near the end of the century.

III. These developments were just consistent with determinism, but the discovery of radioactivity led to the attribution of irreducible randomness to nature.
   A. Radioactivity was discovered in 1896, the same year that the electron and X-rays were discovered, and all three became intertwined.
      1. Radioactivity was discovered accidentally by Henri Becquerel.
      2. Becquerel attended a lecture announcing the discovery of X-rays, repeated Röntgen's experiment, then decided to see if phosphorescent substances gave off these rays naturally.
      3. Fifteen years earlier, Becquerel had prepared a uranium compound for his famous physicist father's research on phosphorescence, and he decided to retrieve that specimen and conduct an investigation.
      4. He stored the uranium, wrapped in a thick black cloth, with a photographic plate and a copper cross; when he developed the plate, the image of the cross was quite clear.
      5. The uranium was emitting rays of some kind but not X-rays.
   B. Becquerel's discovery didn't attract much attention until Marie and Pierre Curie entered the picture.
      1. Marie took up the study of Becquerel's rays in 1897, using an instrument of Pierre's design, and was immediately successful.
      2. In 1898, she established the reality of the phenomenon, named it *radioactivity*, and ranked substances according to their radioactive "power," focusing first on uranium and thorium, then on pitchblende.
      3. Pitchblende led to the discovery of a new radioactive element that Marie named polonium, then to the most powerfully radioactive substance of all, radium, which she isolated in 1902.
   C. The Curies were awarded the Nobel Prize for chemistry in 1903, but by then, radioactivity research had become a very "hot" topic indeed.

1. One important question was: Where did the energy of these radioactive rays come from?
2. Under the influence of the traditional solid concept of the atom, Pierre urged that it came from outside the atom, but in 1899, Marie noted the possibility that it came from a disintegration within the atom involving a loss of mass.
3. Ernest Rutherford, a student of J. J. Thomson's, inclined to an intra-atomic view of radioactivity and became the Curies' greatest research rival.
4. Rutherford and Frederick Soddy discovered that radioactivity is composed of multiple types of rays.
5. They named two of these, alpha and beta, identifying the former as possibly helium nuclei and the latter as electrons; in 1900, Paul Villard identified X-ray-like gamma rays.
6. In 1903, Rutherford and Soddy established that radioactivity was a process internal to the atom, that each radioactive element had a distinctive half-life, that the process was random, and that it involved the transmutation of atoms.
7. Note well: Radioactive decay is a random process, but it is strictly lawful in large populations of atoms, as reflected in the precise, unique half-life assignable to each radioactive element.

D. Rutherford led the way in seeing radioactivity as a "tool" for probing the interior of the atom.
   1. He used a beam of alpha rays to arrive at the *Solar System model* of the atom in 1910, which led Niels Bohr to lay the foundations of quantum mechanics in 1912.
   2. By 1917, Bohr and Einstein showed that the orbital changes of electrons bound to nuclei in the new quantum theory were random, as were the emission and radiation of photons that accompanied these changes.
   3. As we will see, the development of quantum mechanics so anchored the stochastic character of nature that we really did need to rethink what we meant by *knowledge*, *truth*, *law*, and *reality*.

Essential Reading:

Ian Hacking, *The Taming of Chance.*

Theodore M. Porter, *The Rise of Statistical Thinking, 1820–1900*.

Questions to Consider:

1. Does a statistical explanation truly explain, or is it merely descriptive of a state of affairs?
2. How is it that behaviors within large groups are highly predictable even if the behavior of individuals composing the group is random/free?

Lecture Twenty-Seven
Techno-Science Comes of Age

Scope: Although the idea of techno-science originated in the Graeco-Roman period, and technological innovation was a major factor in social change in the 12th century and the Renaissance, it was in the 19th century that techno-science erupted onto the world scene as an omnivorous and relentless driver of social change. Such technologies as the electric telegraph, the transatlantic telegraph, electricity and its host of applications in industry and the home, synthetic dyes and fibers, plastics, artificial fertilizer and hybrid seeds, high-energy explosives, long-distance telephony, radio, and television were made possible by science-informed engineering. And these are merely the tip of the iceberg. By the early 20th century, science, engineering, and innovation had become so intimately intertwined that their respective contributions could not be distinguished. Transportation, communication, construction, production, and information *systems* were all products of this alliance, and all were characterized by continual innovation at an accelerating pace.

Outline

I. In the 19th century, techno-science erupted onto the world scene as a relentless driver of social change.

A. That theoretical knowledge of nature and craft know-how could be integrated fruitfully was already recognized in the Graeco-Roman period.

1. The idea of techno-science is clearly articulated in Vitruvius's *Ten Books on Architecture*, and textual and artifactual evidence shows that mathematical physics was employed technologically to a limited extent from the 3rd century B.C.E. through the 3rd century C.E.
2. The industrial revolution of the 12th and 13th centuries was know-how–driven, thus not an instance of techno-science, but during the Renaissance, mathematical knowledge was made the basis of a wide range of technological innovations.

  3. Furthermore, technological innovation was explicitly identified with progress.
 B. Techno-science was implicit in the creation of modern science.
  1. In the 17th century, Francis Bacon and René Descartes, however different their conceptions concerning knowledge of nature and how to get it, agreed that knowledge of nature would give us power over nature.
  2. It was only in the 19th century, though, concurrent with the creation of new scientific theories of nature, that techno-science went from promise to driver of social change.

II. We tend to deprecate "tinkering," but the first phase of the Industrial Revolution was initiated not by scientists but by tinkerers, gifted with know-how but possessing little or no formal education or knowledge.
 A. Mass production of iron and of textiles, the factory system, and the steam engine were 18th-century inventions that launched the 19th-century Industrial Revolution.
  1. Iron was produced in antiquity, and water-powered blast furnaces were in operation in Europe by the 15th century at the latest, but the scale of iron production skyrocketed beginning in the early 18th century.
  2. Abraham Darby was apprenticed as a teenager to a bronze-smelting master and became a bronze master himself.
  3. He then invented a coke-based process for smelting iron ore in place of the increasingly expensive charcoal-based process then in use.
  4. In 1709, Darby adapted the process to the blast furnace and began commercial production, lowering iron production costs dramatically and increasing production volume many-fold—and at an expanding rate.
  5. Darby also patented a technique for casting iron in molds that enabled mass production of nearly identical pieces, and he incorporated a metallurgy laboratory into his operations, arguably the first industrial research laboratory.
  6. His firm, led by his grandson, built the first iron bridge in 1779, launching iron's use as a structural material.
 B. James Hargreaves and Richard Arkwright, both from desperately poor families, made mass production possible.

1. Hargreaves was a barely literate textile worker whose first *spinning jenny*, developed in 1764, had eight spindles driven by a single operator.
2. By 1777, some 20,000 jennies with up to 80 spindles per wheel were in operation in England alone.
3. Richard Arkwright, taught to read by an older sister, introduced a water-powered *spinning frame* in 1771.
4. In 1775, he patented an improved carding engine to feed the frame; soon, automatic looms were invented to weave the vast output of the growing number of frames into cloth.
5. Unlike Hargreaves, Arkwright became extremely wealthy, largely through his vision of a factory-based system of mass production enabled by the new machinery.
6. As happened with Darby's iron innovations, unit cost (of cotton thread and cloth) collapsed as volume increased a hundredfold and more—but the new factory-based system of production was brutally exploitative of workers, creating wealth for a few from what the Romantic poet William Blake called "dark Satanic mills."

C. James Watt's improved steam engine offered an alternative to water-powered mills, freeing factories from river sites only.
1. Watt was a mechanic/instrument maker who, between 1765 and 1774, developed a much improved version of the original Newcomen steam pump.
2. In 1774, Matthew Boulton bought out Watt's bankrupt backer, immediately adapted Watt's prototype for commercial production, and put Watt to work inventing a string of improved versions.
3. Boulton and Watt steam engines accelerated the Industrial Revolution begun by water-powered mass production, but steam engines soon moved from stationary applications powering mills to mobile applications.
4. The first steam-powered railroads began operation in the 1820s, pioneered by George and Robert Stephenson, then Marc and Isambard Brunel, both father-son teams, with only the last having some formal schooling.
5. Robert Fulton had no technical training at all, but using off-the-shelf engines and hardware and simple arithmetic calculations and with the support of the politically savvy

Robert Livingston, he succeeded in converting his vision of steam-powered boats into commercial reality.

III. In the 19th century, a qualitative change in the nature of technological innovation took place as science became coupled to engineering and entrepreneurship.
  A. Increasingly, invention became one element in a complex process of innovation that was dependent on scientific knowledge, formal engineering training, and supportive business acumen.
  1. The electric telegraph was invented and demonstrated by physicists—Joseph Henry in America and Charles Wheatstone in England—prior to Samuel Morse's efforts, which failed until Joseph Henry was consulted by an associate of Morse and used his physics knowledge to repair the flaw in the design.
  2. Similarly, the first "amateur"-designed undersea telegraph cable, financed by Cyrus Field, failed ignominiously, but its successor, redesigned by the physicist William Thomson, became a global triumph.
  3. Michael Faraday's dynamo of 1830, together with the body of electrical and electromagnetic theory that had grown up in the interim, underlay the emergence of commercial electricity generators from about 1870.
  4. Thomas Edison was a classic know-how genius, but behind his bluster of disrespect for theory, he employed mathematicians, chemists, and physicists to make his inventions commercially successful.
  5. A great deal of scientific and science-based engineering knowledge underlies Edison's scheme for the central generation and distribution of electricity.
  6. Radio technology is directly based on Maxwell's mathematical theory of the electromagnetic field: tinkering without knowledge of electrical theory would not get you very far.
  7. Alexander Graham Bell's telephone worked locally and on a small scale, but more physics was needed to make long-distance work.
  8. In the last decades of the 19th century, the germ theory of disease led to practical applications in vaccination and antisepsis.

B. A particularly good example of the realization of the idea of techno-science is the creation of the synthetic dye industry.
   1. William Perkin made the first synthetic dye, mauve, in 1856 while "tinkering" with coal tar as a course project.
   2. Enormous wealth and power were subsequently created out of chemical knowledge as an engine of technological innovation.
   3. Perkin made a modest fortune, but German chemists generated whole new industries in dyes, synthetic materials, pharmaceuticals, artificial fertilizers, explosives, and more.
   4. All of a sudden, the idea of techno-science became a compelling reality for entrepreneurs, industrialists, financiers, and politicians.

Essential Reading:

Thomas P. Hughes, *American Genesis: A Century of Invention and Technological Enthusiasm.*

Anthony Travis, *The Rainbow Makers.*

Questions to Consider:

1. What is it that drives the pace of innovation—knowledge or desire?
2. Did the very nature of technology change in the 19th century by comparison with the technologies deployed until then?

Lecture Twenty-Eight
Institutions Empower Innovation

Scope: The coupling of newly powerful scientific theories to invention changed the very nature of technological innovation, opening the door to world- and life-transforming technologies. To enter that door, Western societies created institutions explicitly designed to promote innovation, to "capture" it, and to channel it into society. It was this institutionalization of innovation, doubtless provoked by the power of new inventions and the prospect of unprecedented profitability, that made the social impact of technology since 1800 different from its impact in the Graeco-Roman and medieval periods. It also explains why the pace of innovation since 1800 has been so great: These institutions demand continuous innovation. The enablers of modern techno-science as an agent of change include science-based engineering education serving innovation-driven industrial corporations, the university as a generator of new knowledge, commercial institutions keyed to innovation-driven economic growth, and governmental policies linking security and prosperity to continuous innovation.

Outline

I. The institutionalization of innovation was the key to the social impact of modern technology.
   A. By the early 19th century, all the pieces were in place for transforming the scale and scope of technological innovation.
      1. All technologies may be artificial, but they are not all equal from a societal perspective.
      2. In the course of the 19th century, technological innovation changed its character and acquired new social power.
      3. Instead of building on or enhancing existing capabilities, techno-scientific innovations introduced capabilities that never existed before.
   B. How this transformation took place such that it has been sustained for 200 years is not self-evident.

1. Invention by itself cannot explain the transformation because it is episodic, while 19th-century techno-science became a stable, almost continuous source of repetitive innovation.
   2. Entrepreneurialism cannot explain the transformation because it is too narrowly motivated to be the driver of social change that techno-science became.
  C. One important factor was a growing perception—which appeared just as the Industrial Revolution was getting underway—that invention and scientific knowledge could be coupled to create a true engine of economic growth.
   1. The perception appears in Adam Smith's *The Wealth of Nations* [with its chapters on how division of labor can maximize the efficiency of factory-based mass production].
   2. It appears in Alexander Hamilton's *Report on Manufactures* to the new American Congress, in which he championed industrialization over Jeffersonian gentleman farming as the basis for the new republic's future.
   3. This debate raged throughout the 19th century in the United States in different sectors of the still expanding Union, but recognition grew that institutions can be created and reformed so as to encourage technological, knowledge-based innovation that can be channeled into society.
II. The key to unleashing the power of techno-science in the 19th century was institutionalization.
  A. A deliberate effort was made to create new institutions—educational, business, and governmental—that would systematically couple science and engineering to create a sustained engine of growth based on technological innovation.
   1. One globally influential outcome of the French Revolution was the creation of the École Polytechnique, the first modern engineering school.
   2. It is modern in the sense that it was the first school in which the engineering curriculum was based on rigorous training in mathematics and physical science and in the laboratory.
   3. The curriculum was, in fact, intended to train engineers in techno-science.
   4. The École continues to exist, and an astonishing number of influential French figures continue to be among its graduates.

        5. Henri Becquerel, for example, was a fourth-generation École alumnus.
- **B.** The École curriculum was a major international influence, but it also precipitated bitter controversies in educational circles that lasted for decades.
    1. That engineering education should be based on science, math, and laboratory work rather than machine shop and field experience was highly controversial in many circles.
    2. That engineering education should be considered "higher" education at all was highly controversial, especially at elite universities that were suspicious of teaching science, let alone engineering.
    3. Ironically, the British, whose inventive "tinkerers" and deep pool of mechanics gifted with creative know-how had started the Industrial Revolution, strongly resisted reforming engineering education as a college/university curriculum.
    4. This reluctance came in spite of the fact that by 1850, Parliament was warned that Germany and America would overtake England as an industrial power if engineering education were not reformed.
    5. Late in the century, engineering training in England still required an apprenticeship, as if engineering were a craft guild.
    6. An ironic consequence of this was that J. J. Thomson, who would win a Nobel Prize in physics and whose students, Ernest Rutherford and Francis Aston, also won Nobel Prizes, went to college only because his family could not afford the engineering apprenticeship fees!
- **C.** Engineering education in the United States also struggled mightily to establish itself as a science-math-lab–based university-level curriculum.
    1. West Point Military Academy adopted the École curriculum from its inception, and that curriculum evolved into the most generally used curriculum in the United States, though it was supplemented by a good deal of machine shop and field work.
    2. Until 1862, there were literally only a handful of engineering colleges in the United States, but passage of the Morrill Land Grant Act that year created very substantial financial incentives for states to establish engineering colleges.

3. After the Civil War, U.S. engineering colleges grew explosively, as did the engineering student population.
4. The number of U.S. engineers doubled every decade from 1880 to 1960, reaching some 2 million by 1980.

D. This explosive growth is a social phenomenon, not an educational one.
1. Institutions needed to be created to educate the growing numbers of students who wanted to study engineering.
2. Not only is creating such a foundation capital intensive, but given an École-style curriculum, it also requires a massive increase in the number of physicists, chemists, and mathematicians, as well engineering faculty.
3. This is perhaps the single most important cause of the existence of a U.S. community of research-oriented physicists, chemists, and mathematicians in the 20th century.
4. The increasing number of engineering students was itself the effect of a transformed U.S. industry that required science-trained engineers to ride the techno-science bandwagon.

E. The positive feedback relationship between engineering and industry made people aware of the social consequences of riding this bandwagon.
1. The creation of the Georgia Institute of Technology in the 1890s became a battleground within southern society over the expanded industrialization implicit in preparing so many engineers for jobs that required techno-science training.
2. The battle was over the nature of the curriculum, and initially, approval to build Georgia Tech was contingent on a machine shop-based curriculum, a startling contrast from MIT and Cal Tech, for example.
3. By the 1890s, the character of U.S. science and engineering education had been enriched by the German research model and reflected the latest developments in physics, chemistry, and mathematics and their industrial applications.

III. Demand for such educational reforms came from new business and governmental institutions that made technological innovation fundamental to commercial success and national power.

A. Francis Lowell's creation of an integrated textile factory put U.S. industry on the path to the modern industrial corporation.

1. Lowell's idea led to a host of innovations in production machinery, in water-power generation and transportation technologies, and in business organization and workforce management.
2. The competitive advantage of the integrated enterprise was a reduction in transaction costs and time, and as the value of time increased, so did the value of integrating firms to reduce transaction costs.
3. The social impact of aggressively competitive integrated firms was profound, and sustained competitive success demanded continuous growth through repeated innovation.

B. The Swift Meat Packing Company illustrates the technology-intensive character of the integrated modern industrial corporation and the vast social "ripples" it generated.
1. Swift created a competitive juggernaut and only by copying him could his competitors survive.
2. The complexity of Swift's business model is not widely appreciated, but it should be.
3. Those competitors who emulated his model, such as Armour, joined with Swift to form a national "meat trust" that in 1905 was declared by the Supreme Court to be in violation of antitrust laws created to protect society against unfair competitive advantage.
4. But Swift's institutional model was a universal competitive machine, and entrepreneurs across all industries saw the competitive advantage of emulating that machine.
5. Standard Oil, General Electric, U.S. Steel, Westinghouse, Ford, DuPont, and General Motors, among many others, were all organized along these lines.

C. New government institutions that were keyed to techno-science also needed to be created.
1. Except for a contentious alliance during World War I, the U.S. government kept direct-support scientific research to a minimum and provided barely indirect support for technological innovation.
2. World War II changed all that with the creation and extraordinary success of the Office of Scientific Research and Development, headed by Vannevar Bush.

3. At the end of the war, Bush presented President Truman with a report, *Science: The Endless Frontier*, that, together with the Cold War, created a mandate for large-scale federal support of scientific and technological research and development.
4. It launched a raft of government institutional reforms that have woven science and technology more deeply into the fabric of modern life.

Essential Reading:

Thomas P. Hughes, *Networks of Power*.

G. Pascal Zachary, *Endless Frontier*.

Questions to Consider:

1. Since the rise of modern techno-science, science and science-based engineering certainly have *changed* the conditions of human life, but is life *better* for those changes? For whom and at what cost?
2. Where does responsibility for the social and physical impacts of innovations lie: in knowledge, in technologies, in society?

Lecture Twenty-Nine
The Quantum Revolution

Scope: Quantum physics is without question the most esoteric of all scientific theories, and the most inaccessible without advanced training in that theory. It is the most revolutionary of 20th-century theories, replacing wholesale the inherited conceptual framework of modern physics. At the same time, quantum mechanics is arguably the most important physical scientific theory of the 20th century and the most explanatorily powerful, experimentally confirmed, and predictively successful physical theory ever. In spite of this, and in spite of the wealth of its practical applications, the interpretation of the theory has been controversial since the mid-1920s, and it is inconsistent with the general theory of relativity. Quantum mechanics imputes randomness, probability, and uncertainty to elementary physical processes and to the elementary structure of physical reality. It redefines causality, space, time, matter, energy, the nature of scientific law and explanation, and the relationship between mind and world. It is exhilarating!

Outline

I. The quantum idea seems to be the most bizarre idea in modern science, but it became the cornerstone of a powerful theory of matter and energy.

   A. Initially, the quantum idea was that electromagnetic energy only comes "packaged" in discrete sizes.
      1. In 1900, Maxwell's theory of electromagnetic energy and the electromagnetic field was at the heart of theoretical physics.
      2. Some argued that electromagnetic energy was the ultimate stuff of physical reality and that matter was reducible to it.
      3. The concern of physicists with a correct theory of the aether was motivated by the need to anchor the physical reality of the electromagnetic field.
      4. A fundamental feature of Maxwell's theory is that electromagnetic energy is continuous.

B. The story of quantum *mechanics* begins with Niels Bohr, but the quantum *idea* preceded Bohr.
   1. Max Planck invented the quantum idea in late 1900 as a way of solving a gnawing problem in the physics of the day.
   2. This problem required for its solution a description of how the electromagnetic energy of an ideal *black body*—one that absorbed all the radiation that fell on it—was distributed as a function of frequency.
   3. This seemed a simple problem, but it resisted a comprehensive solution until December of 1900, when Planck solved it. He was displeased with the assumption his solution required, however: that radiation is emitted/absorbed in quantized units.
   4. The electromagnetic field is continuous and electromagnetic waves are continuous; thus, electromagnetic energy should be continuous, but somehow the process of emission/absorption *seems* discontinuous.

C. While Planck was trying to kill the quantum idea, Einstein made it real.
   1. In 1905, Einstein showed that the quantum idea, if treated as real, could explain the puzzling photoelectric effect.
   2. Einstein was awarded the Nobel Prize for his paper on this, but note that it overturns the wave theory of light and replaces it with an "atomic" theory of light.
   3. These quanta of light, paradoxically, also have a frequency and, thus, must in some sense be wavelike, and their energy follows Planck's formula of 1900.
   4. Einstein treated the quantum idea as a new fact about physical reality and, from 1906 to 1908, used it to solve other problems in physics and chemistry (though as he focused on the general theory, he abandoned atomic units of reality!).

D. In 1910, Ernest Rutherford proposed a Solar System model for the atom, but it was seriously flawed: electrons should immediately spiral into a positively charged nucleus, according to Maxwell's theory, yet atoms were stable!

E. Niels Bohr worked in Rutherford's laboratory in 1911 and rescued Rutherford's model by proposing new principles of nature at the subatomic level, as if Maxwell's theory were not allowed in there.
   1. Orbital electrons do not radiate energy while orbiting; they radiate energy whenever they change orbits.

2. For each chemical element, the electrons in its atoms are strictly limited to a discrete set of "permitted" orbital radii only, each associated with a specific quantized energy level.
3. The energy radiated/absorbed when electrons change orbits is equal to the difference between the energy level of the orbit from which the electron starts and that of the orbit at which it arrives.

F. These are, on the face of it, bizarre assumptions and they violate Maxwell's theory, but there was a powerful payoff.
1. Since the mid-19th century, a mountain of data had accumulated by spectroscopists documenting distinctive frequencies of light emitted by each element without explaining this phenomenon.
2. Bohr predicted that these frequencies corresponded precisely to "permitted" orbital transitions for an atom of each element and showed that this was indeed the case for hydrogen.
3. Further, Bohr predicted that the periodic table of the elements, another unexplained "fact" about matter, was a consequence of the way atoms were built up as electrons and protons were added to the "basic" hydrogen atom.

II. Building quantum mechanics on the quantum idea took place in three distinct stages and has had profound social consequences through its applications.

A. The first stage extended from Bohr's adoption of the quantum idea in 1912 to 1925.
1. Bohr's quantum rules for orbital electrons attracted a cadre of physicists who, over the next 10 years, dramatically extended the quantum theory of the internal structure of the atom and fulfilled Bohr's vision of explaining the periodic table.
2. Ironically, the theory stimulated new and much more precise spectroscopic experimentation that, by the early 1920s, had Bohr's initial quantum theory of the atoms on the ropes!
3. The new spectroscopic data could not be explained by Bohr's original version of quantum theory.
4. This precipitated a genuine crisis, with some suggesting that the quantum idea should be abandoned entirely.
5. Concurrently, Louis de Broglie argued for extending the quantum idea from radiation to matter, in particular, the

coexistence of wave and particle features Einstein had proposed for quanta of radiation.

B. The second stage extended from the resolution of the crisis in 1925 to the early 1960s.
   1. Werner Heisenberg and Erwin Schrödinger independently resolved the crisis in 1925 by creating quantum mechanics, though their respective formulations were profoundly different from each other.
   2. Between 1926 and 1929, Heisenberg saw that his quantum mechanics implied an *uncertainty principle*, and he and Bohr developed a statistical interpretation of quantum mechanics and a radically new interpretation of physical theory based on it.
   3. In 1929, Paul Dirac published a new theory of the electron that evolved into a quantum-level analogue of Maxwell's theory, called *quantum electrodynamics* (QED).
   4. Quantum physics also provided a theoretical framework for designing increasingly powerful particle accelerators from the early 1930s to the present.

C. The third stage extended from the early 1960s to the present as QED evolved into *quantum chromodynamics* (QCD).
   1. By the early 1960s, particle accelerators had created some 200 "elementary" subatomic particles, which was clearly impossible!
   2. In 1964, Murray Gell-Mann and George Zweig independently proposed a new theory of matter, later called QCD, which is a theory of the so-called strong force holding the nucleus together.
   3. In QCD, the elementary building blocks of all nuclear particles are quarks. [Quarks are held together by massless particles called gluons, while electrons are members of a family of elementary particles called leptons.]
   4. As QCD developed, the number of quarks needed to build up all known particles reached six. [QCD also included a matching number of leptons and eight types of gluons.]

III. Also in the early 1960s, physicists began the pursuit, still unachieved, of a unified theory of the four known forces in nature.

A. The four forces are: electromagnetic, weak (affecting leptons), strong (affecting quark-based particles), and gravity.
1. The first step toward unification of these forces was to unify the electromagnetic and weak forces.
2. Based on a suggestion by Sheldon Glashow for how this could be accomplished, Abdus Salam, Steven Weinberg, and Glashow succeeded in developing an electro-weak theory.
3. The theory predicted the existence of a hitherto unsuspected family of particles, and experiment soon confirmed their existence with the properties predicted by the theory.

B. The next step was unifying electro-weak theory and QCD, the theory of the strong force.
1. Actually, unification of these two theories overlapped and was accomplished by the late 1980s.
2. This so-called *standard model*, although not without some problems, has achieved impressive experimental confirmation.
3. It also works at levels ranging from the subatomic to the cosmological.
4. The final step, uniting the standard model with the gravitational force into a *theory of everything*, has so far proven elusive.

Essential Reading:

George Gamow, *Thirty Years That Shook Physics*.

Steven Weinberg, *Dreams of a Final Theory*.

Questions to Consider:

1. How could a theory as tested as quantum theory, and with so many complex technologies based on it, possibly be wrong?
2. What is energy "made of" that the Universe can be described as arising out of negative quantum vacuum energy?

Lecture Thirty
Relativity Redefines Space and Time

Scope: Between 1905 and 1915, Einstein effectively redirected the course of modern physics. Although he is identified with the special and general theories of relativity, he was also a leading architect of quantum theory up to the formulation of quantum mechanics in 1925. The special theory solved a problem central to 19th-century physics that had resisted solution for decades. In the process, it eliminated the aether (and a generation of aether physics), forced a reconceptualization of Newtonian space and time, and proclaimed the interconvertibility of matter and energy. The general theory went much further. It was a universal theory of gravity that redefined physical reality at the cosmological level. Space and time were dynamically interrelated with matter and energy, their properties determined by the evolving distribution of matter and energy. Both theories made utterly unanticipated predictions that have been confirmed, especially in the 1930s and since the 1960s.

Outline

I. Einstein's special and general theories of relativity, introduced in 1905 and 1915, respectively, have had a profound impact on our conception of reality.

  A. Independently of relativity theory, two other papers written by Einstein in 1905 had already altered our conception of reality.

  B. Einstein published a theory of Brownian motion that played a major role in convincing physicists of the reality of atoms, that matter really was discontinuously distributed in space.

  C. Einstein also published his theory of the photoelectric effect we discussed in the last lecture, which made quanta of radiation (photons) real, overthrew the wave theory of light, and introduced the seemingly dual nature of these quanta, later extended to matter.

II. As if that were not quite enough for one year, in 1905 Einstein also published a paper called "On the Electrodynamics of Moving Bodies."

  A. This is the paper that founded the special theory of relativity (STR).

1. STR dismissed the reality of the aether, which had become central to physicists' account of reality from the 1840s, as a mistake.
2. The mistake was not taking into account how we measure time and space, using clocks and rulers, and the role that light signals play in making these measurements.
3. Einstein postulated as a law of nature that the speed of light is a constant for all observers regardless of their motion relative to that light.
4. Numerous 19th-century experiments suggested this, but it is a conceptually bold and somewhat bizarre idea.

B. Einstein also postulated a principle of relativity that physicists had been using since Galileo and Huygens, namely, that the laws of nature are indifferent to uniform (unaccelerated and, hence, force-free) motion.

1. Thus, two observers moving uniformly with respect to each other and doing the same experiment would "discover" the same law of nature at work in that experiment.
2. Einstein showed that combining this long-established principle with the constancy of the speed of light and applying both systematically allows us to account for all known mechanical, electromagnetic, and optical phenomena, without any need for the aether.
3. The aether, Einstein claimed, was invented by physicists to explain a problem that disappears when one accepts the constancy of the speed of light and applies the principle of relativity in electromagnetic theory, as well as in mechanics.
4. The constant, finite velocity means that the time it takes for light signals to give observers information from rulers and clocks affects the results of those measurement operations.
5. The results of measuring spatial intervals and temporal intervals depend on the relative motion of the observer and what he or she is measuring.
6. Two observers moving relative to each other and measuring the same lengths and times will get two different answers, but these are strictly correlated by the laws of STR.
7. Each can predict the other's results once the relative motion of each is known.

- **C.** STR implies a profound change in our definitions of space and time, matter and energy.
  1. We have discussed the historic roots of the principles of the conservation of matter and the conservation of energy.
  2. The equations of STR have as a logical consequence that matter and energy are not, ultimately, two different features of reality, and they are not conserved independent of each other.
  3. The equation $E = mc^2$, unsought for in advance, predicts that they are interconvertible and only jointly conserved.
  4. Einstein suggested that this surprising connection between matter and energy could be tested by extending Marie Curie's experiments on radioactivity.
- **D.** Since Newton, space and time had an absolute character, as if they were things with properties of their own independent of what happens in space and time.
  1. In STR, talk of space and time in physics is meaningful only as talk of measurements relative to particular frames of reference.
  2. Absolute space and time are replaced by local space and local time, but all possible "locales" are strictly correlated with one another by the equations of STR.
  3. This is a fundamental point about STR and the general theory of relativity: They are strictly deterministic theories!
  4. What is relative in STR is not reality but the measurement of lengths and times: The results of measurements are relative to the frame of reference of the observer, but all possible observers are correlated deterministically.
  5. Quantum theory, to which Einstein made fundamental contributions from 1905 to 1925, evolved into a stochastic theory of nature, and for that reason, Einstein never accepted its finality as a theory. He was irrevocably committed to determinism as the only rational account of experience.

III. The general theory of relativity (GTR) appeared in 1915, after eight years of intense work that Einstein said sometimes made him feel he was going mad.

- **A.** GTR removes the limitation in STR of the principle of relativity to observers in uniform motion.

1. As with STR, Einstein begins by postulating as a principle of nature something physicists had known about for centuries but whose implications had been ignored.
2. This principle was the equivalence of gravitational mass (weight in a gravitational field) and inertial mass (the force required to accelerate an object, even if it is "weightless," as a satellite in orbit is).
3. This equivalence implies that there should be no difference in the laws of physics for an observer stationary in a gravitational field or experiencing a force independent of a gravitational field.

B. When the laws of physics are reformulated in a way that reflects this property of invariance, reality changes!
1. For one thing, space and time in GTR really are names of relationships, not things in their own right.
2. Further, there is no way to specify the properties of space and time independently of what happens in space and time.
3. There is, for example, no universal geometry to space, contrary to the age-old assumption that Euclidean geometry was the absolute background geometry for all of nature.
4. The geometry of space is determined by the distribution of matter and energy within space: "flat"/Euclidean where the density of matter/energy is very, very small and "curved"/non-Euclidean where the density is greater, with a radius of curvature that is a function of the density.

C. GTR as a theory of space, time and gravity is independent of quantum physics as a theory of the natures of matter and energy.
1. GTR is a theory of the Universe as a whole, a cosmological theory.
2. Einstein fine-tuned the initial version of the theory to reflect the prevailing conviction that the Universe as a whole was static.
3. From 1917 to 1923, however, physicists found that GTR was a dynamic theory that predicted an expanding Universe.
4. Einstein introduced a term into the GTR field equations that cancelled the expansion, but in 1929, Edwin Hubble announced that the Universe *was* expanding!

D. Like STR, GTR has withstood intense experimental testing and has made startling predictions that have been confirmed.

1. Most recently, astronomers have confirmed that GTR correctly predicted the observed orbits of two pulsars around each other.
2. NASA's *Gravity Probe B* confirmed the prediction that the rotational motion of the Earth twists spacetime in its immediate vicinity.
3. Apart from being used to refine the accuracy of the Global Positioning System (GPS), GTR has had no other technological implications that I know of, but GTR has wholly transformed our most fundamental scientific ideas about the Universe and its structure.
4. STR, by contrast, is familiar from its application to atomic energy (nuclear and thermonuclear technologies), and as a feature of the design of particle accelerators, but less familiar for its assimilation into explanation in physics at the deepest level.

Essential Reading:

Albert Einstein, *Relativity: The Special and General Theory*.

John Stachel, *Einstein's Miraculous Year*.

Questions to Consider:

1. What do Einstein's theories tell us about the power of the human mind to discover the truth about the world "out there" just through thinking?
2. Why was Einstein so committed to determinism?

Lecture Thirty-One
Reconceiving the Universe, Again

Scope: The Newtonian Universe was spatially infinite and materially finite. The consensus view, although it was not unanimous, was that the material Universe was identical with the Milky Way, whose size was first determined during World War I. In the 1920s, the scale of the Universe changed dramatically with the discovery of thousands of galaxies extending to the limits of observability. Then, Edwin Hubble announced that the Universe was expanding. Together with the general theory of relativity, cosmological speculation became a branch of science and made a scientific problem out of the origin of the Universe. After World War II, the Big Bang and Steady-State theories were offered as rival explanations, even as astronomers were discovering a rich, complex, invisible Universe behind the visible. In the 1980s, the scale and the nature of the Universe changed again, the vastness now beyond imagination, and the detectable Universe, visible and invisible, a mere footnote to a newly conceived whole.

Outline

I. The Universe was reinvented in the aftermath of Copernicus's theory of the heavens and refined in the 18th century, then reinvented again in the 20th.

    A. Copernicus opened a door and philosophers and scientists streamed through it.
        1. The Universe as Copernicus himself conceived it was superficially similar to the medieval Universe.
        2. Both were spherical and, hence, had a well-defined center and periphery, though Copernicus's was much larger.
        3. Copernicus's Universe was heliocentric, not geocentric, and this stripped it of all explicit symbolic meaning.
        4. De-centering the Earth naturalized and secularized the Universe.

    B. The naturalistic implications of Copernicus's idea were quickly extended by others.

1. Thomas Digges adopted Copernicus's idea but made the Universe infinite in space, populated by an endless number of stars.
2. Copernicus and the infinity of the Universe were championed by Giordano Bruno in his *Infinite Universe and Worlds*.

C. By the end of the 18th century, Copernicus's ideas, amended by Kepler and Newton, were triumphant: The Universe was wholly natural and infinite.
1. Consider the title of Thomas Wright's 1750 book: *An Original Theory or New Hypothesis of the Universe, Founded upon the Laws of Nature, and Solving by Mathematical Principles the General Phenomena of the Visible Creation*….
2. Immanuel Kant was inspired by a review of Wright's book to formulate a Newtonian cosmological theory of his own in his *Universal Natural History and Theory of the Heavens*.
3. Kant's Universe is a boundless, gravitationally ordered hierarchy of systems of galaxies composed of solar systems like our own, continually forming, collapsing, and reforming out of gaseous nebulae, according to Newtonian laws.
4. Both of these were superseded in fame and impact by Laplace's 1796 *Exposition of the System of the World*, followed by his *Celestial Mechanics*.

II. The scientific study of the "heavens" and their glory was part of the progressive naturalization by scientists of the Earth, life, mankind, and mind.

A. The consensus scientific view of the Universe was that it was spatially infinite but contained only one galaxy, the Milky Way.
1. The scale of the Solar System was well known in the 18th century, but the actual distance to the stars was first measured in 1838.
2. From its parallactic displacement at opposite sides of the Earth's orbit, Friedrich Bessel determined the distance to the star Cygnus 61 to be about 10 light years from Earth.
3. This led to crude guesses at the size of the Milky Way, then considered to *be* the Universe, but in 1912, Henrietta Swan Leavitt invented a cosmic "ruler" based on variable stars.
4. Harlow Shapley used Leavitt's ruler to determine that the Magellanic Clouds were some 400,000 light years away from

Earth and that our Sun was 50,000 light years from the center of the Milky Way, which was, thus, more than 100,000 light years across.

B. In 1920, Harlow Shapley and Heber Curtis debated the question of whether the Milky Way was the Universe.
  1. Shapley argued that it was and thought he had won the debate, but even if he did, he was wrong about the Universe!
  2. Shapley left Mt. Wilson Observatory for the directorship of the Harvard Observatory just as a new 100-inch telescope was becoming operational at Mt. Wilson.
  3. The new Mt. Wilson director, Edwin Hubble, announced in 1923 that Andromeda was a galaxy, not a glowing cloud of gas within the Milky Way, as Shapley thought.
  4. Hubble estimated its distance at a million light years and soon announced that there were thousands of galaxies, extending as far as the telescope "eye" could see.

C. All by itself, this transformed the vastness and complexity of the Universe, but it was just the beginning.
  1. In 1929, Hubble announced that the Universe was not static—it was expanding.
  2. Hubble based this claim on his analysis of stellar and galactic *red shifts*, which implied that effectively all stars and galaxies were racing away from the Earth.
  3. The red shift is an optical version of the Doppler effect, more familiar in its acoustic form as the effect of motion on the perceived frequency of a sound.
  4. This shift had been observed for a dozen or so stars around 1912 by Wiley Slipher, who published this observation, but it was ignored.
  5. Hubble made thousands of such measurements and incorporated them into an interpretation of the Universe as expanding.

D. Of course, viewing the expansion of the Universe in reverse has a conceptually dramatic consequence.
  1. An expanding Universe suggestively started from an initial first moment.
  2. For some astronomers, this initial moment was the "birth" of the Universe: In other words, the Universe had a beginning in time.

III. The expansion of the Universe needed explaining, but the newly vast Universe also raised the question of what was out there.
  A. The Universe was, it turned out, not just the things we could see: It was filled with things we could detect but not see.
   1. In the mid-1930s, radio engineer Karl Jansky accidentally discovered that there were electromagnetic sources in the sky, giving rise to the new discipline of radiotelescopy.
   2. Radiotelescopy revealed that "empty" space was not empty, that molecules and clouds of molecules, some complex, were everywhere.
   3. Concurrently, cosmic rays were recognized as subatomic particles "raining" on the Earth from outer space.
   4. Over the next decades, X-ray, gamma-ray, neutrino, and gravity-wave telescopes were built.
  B. In 1980, astronomers—trying to answer a longstanding question concerning why galaxies are so stable—proposed that galaxies were surrounded by a "halo" of a new form of matter they called *dark matter*, which made up the bulk of the matter in the Universe.
  C. In 1998, observations suggested that the expansion of the Universe was accelerating, not slowing down, as expected. The cause was *dark energy*, which made up the bulk of the matter-energy of the Universe, with dark matter in second place and "our" kind of matter just about 4 percent of the whole.
  D. Furthermore, back in 1980, physicist Alan Guth had proposed an *inflationary theory* of the birth of the Universe that meant the Universe was unimaginably vaster than what we could detect.

IV. The problem of the expansion of the Universe raised the question: Did the expansion imply a beginning, or has the Universe been expanding forever?
  A. Shortly after World War II, George Gamow and Fred Hoyle, with collaborators, proposed rival responses to this question.
   1. Gamow proposed what Hoyle mockingly dubbed the *Big Bang theory* of the Universe.
   2. Hoyle proposed a *Steady-State theory* of an eternally expanding Universe.
   3. The Big Bang theory predicted a universal, uniform microwave background radiation, but it could not explain how

elements heavier than lithium could be built by stars out of hydrogen.
   4. Hoyle, too, needed such a nucleosynthesis theory, and he got one out of William Fowler and Margaret and Geoffrey Burbidge.
   5. The discovery in 1963 of the microwave background radiation that Gamow had predicted shifted opinion to the Big Bang theory, now incorporating the Hoyle-inspired solution to the nucleosynthesis problem!
B. Quantum theory was needed to solve both the nucleosynthesis problem and the problem of the origin of the Universe.
   1. Explaining our Universe using quantum theory in the 1970s solved some problems but created others, including the failure to observe magnetic monopoles.
   2. In 1980, Alan Guth proposed that the Universe began with an instantaneous "inflation" of a primordial dot of negative quantum energy.
   3. This inflation was by a factor of 2^{100}, and detailed observations of the microwave background radiation support its reality.
   4. A corollary of this inflation is that everything we previously meant by *Universe* is only a minute patch on a vastly greater "true" Universe that is forever beyond interaction with us.

Essential Reading:

Brian Greene, *The Fabric of the Cosmos*.

Dennis Overbye, *Lonely Hearts of the Cosmos*.

Questions to Consider:

1. How do we know that the only cause of the red shift is motion?
2. Why is modern cosmology less shocking to people than Copernicus's theory was?

Lecture Thirty-Two
The Idea behind the Computer

Scope: J. David Bolter introduced the idea of a "defining technology" of an era or a society: for example, the mechanical clock, the steam engine, electricity, and the automobile. The computer is a defining technology today, yet in spite of computers being ubiquitous, the idea underlying the computer and giving it its power is not widely appreciated. The modern electronic digital computer has roots in the pursuit of increasingly powerful automatic arithmetic calculators and algebraic problem-solvers and in the design of automated information tabulation. But the modern computer is not a calculator or a tabulator. The idea underlying the computer derives from the solution by Alan Turing of a highly abstract problem in the foundations of mathematics. Turing imagined a machine that could solve any problem whose solution could be specified by a finite decision procedure, or algorithm. Turing and, independently, John von Neumann recognized that emerging new calculators could be reconceived as generalized problem-solving machines, even artificially intelligent machines.

Outline

I. The computer is the "defining technology" of our era, though *semiconductor-based microelectronics technologies* would be a more accurate term.

  A. The computer is ubiquitous today, but just what is a computer and what is the idea of the computer?
  1. Through World War II, *computer* was a job description for someone hired to compute, to make calculations.
  2. Scientists, mathematicians, and users of mathematics in commerce, industry, and the military require the results of vast numbers of tedious, repetitive calculations.
  3. The expense, time, and error-rich character of producing these results provoked Charles Babbage to invent his Difference Engine in 1822 to produce them automatically.
  4. While attempting to build a prototype Difference Engine, Babbage designed his Analytical Engine, a programmable,

stored-memory calculator that could solve a much wider range of mathematical problems.
    5. Neither engine was built, but the latter is often called the forerunner of what we mean by *computer*.
  B. The increasing size and complexity of 19th-century businesses also increased the volume of "number crunching" routinely required to manage businesses effectively.
    1. Mechanical calculators became widespread in the course of the century, beginning with the Arithmometer, based on Leibniz's design and, from 1887, keyboard entry machines such as those offered by Burroughs.
    2. Adding machines became ubiquitous, but the demand for more numbers fed on itself.
    3. In the 1930s, *super-calculators* appeared as military, industrial, commercial, and research needs outran the effectiveness of both human computers and arithmetic calculators.
    4. In 1936, Konrad Zuse applied for a patent on a programmable, stored-memory, stored-program, binary-base and floating-point arithmetic automatic calculator using mechanical relays and built a series of functional machines: the Z1, Z3, and Z4.
    5. Concurrently and independently, George Stibitz at Bell Labs and Howard Aiken at Harvard (with major funding from IBM) designed and built electromechanical, relay-based, programmable super-calculators.

II. The demand for calculation drove the mechanization of arithmetic operations, but calculators are *not* defining-technology computers.
  A. Calculator technology surely would have followed a path of development analogous to that followed by the "true" computer.
    1. Electromechanical machines, however massive, are slow and limited, and the switch to electronic technology had already begun before World War II.
    2. With the invention of the transistor in 1947, totally independently of calculators and computers, calculators migrated to semiconductor technology.
    3. This put them on the path to become ever smaller, more powerful, and cheaper devices.

B. Through the convergence of quite independent lines of development, the "true" computer emerged.
  1. Between 1927 and 1942, Vannevar Bush was involved in the design and construction of a series of "analog computers" for solving engineering problems posed by the electrical power system.
  2. His 1931 Differential Analyzer was especially successful and influential, both in the United States and England.
  3. John Atanasoff and Clifford Berry, impressed by Bush's machine, built a digital, *electronic* computer to solve large numbers of simultaneous linear algebraic equations, a very important problem-solving capability for science and engineering.
  4. This machine used 300 vacuum tubes, was programmable, had storage capability (using condensers), used binary and floating-point arithmetic, and had an automatic printer.
  5. In 1943, John Mauchly and John Eckert, familiar with the Atanasoff and Berry computer, received funding to build ENIAC, initially to produce ballistics tables for the military, but soon perceived to be a general-purpose mathematical problem-solver.
  6. ENIAC was a colossus, 100 feet long with 18,000 vacuum tubes and drawing 100 kilowatts of electrical power, and it was successful, though only operational after the war was over.

III. ENIAC was a watershed machine, dividing calculating from information processing.
  A. The crucial piece in the story of the true computer begins with an esoteric piece of abstract mathematics.
   1. In the mid-1930s, Alan Turing solved an important problem in the foundations of mathematics that had been set by David Hilbert in 1900.
   2. Turing proved that the problem could not be solved, that there could be no finite, "mechanical" decision process that would guarantee the correct solution of all mathematical problems.
   3. A byproduct of this proof was a machine Turing imagined that *could*, in a finite number of steps, solve any problem for which such a decision process, an *algorithm* as it came to be called, *could* be specified.

4. Turing spent a year at Princeton, where he met John von Neumann, head of the new Institute for Advanced Studies.
5. As a wartime consultant to the U.S. government, von Neumann coincidentally learned of the ENIAC project and recognized the potential for an electronic digital computer to be such a Turing machine.
6. He began the design of EDVAC, an enhanced ENIAC, in 1943 and built it at Princeton in 1948.
7. This was a realization of Turing's imagined machine in electronic form, and von Neumann laid down the basic rules for designing and programming general-purpose problem-solving computers that remain dominant.

B. The evolution of the computer has been enabled by the coordinate evolution of microelectronics technologies, especially silicon semiconductor technologies.
1. The integrated circuit was developed independently in 1958 by Jack Kilby and by Robert Noyce, and the silicon chip evolved out of that initially crude device.
2. In 1965, Gordon Moore observed that the number of transistors on a single silicon chip seemed to be doubling every year, implying that by 1975, this number would reach 64,000, which he thought likely.
3. Today, there are about a billion transistors on a single chip, just about keeping up with Moore's "law" (which is not a law of nature at all, of course), but in the near future, current technology will be undermined by quantum effects as chip features approach atomic dimensions.
4. The importance of this continual increase in the number of transistors on a chip lies in the increasing speed with which that chip can execute the instructions it is given.

C. Here, finally, is the essence of the computer as the defining technology of our age.
1. The computer is not a calculator but a universal *simulator*.
2. Nothing that goes on inside the logic or memory circuits of computer chips corresponds to anything whatsoever of interest to us, not even arithmetic calculations.
3. When a computer actually is used to make calculations, that, too, is a simulation.

4. What goes on inside computer chips is as follows: Some combination of high- or low-voltage electrical signals on the input pins of the chip package are transformed into a different combination of high- or low-voltage signals on the output pins of the package in accordance with a set of transformation rules that are wired into that chip.

D. Everything useful a computer does—calculation, word processing, running video games, *everything*—is a simulation based on software.
1. A computer program is a systematic correlation of the electrical signal transformation rules wired into a chip with something meaningful to us: music, video, letters on a screen, and so on.
2. Every program is an algorithm, a set of instructions that completely specifies a finite set of electrical signal transformations that have a symbolic interpretation given by the program.
3. The power of the computer lies in its ability, foreseen by Turing, to generate what its user can interpret as a solution to any problem for which the user can provide such an algorithm.
4. Behind the power of the idea of universal simulation lies another powerful scientific idea: information.

Essential Reading:

Paul Ceruzzi, *A History of Modern Computing*.
Andrew Hodges, *Alan Turing: The Enigma*.

Questions to Consider:

1. Is the substitution of simulation for reality a threat to our humanity?
2. What are the implications of the total artificiality of computer *science*?

Lecture Thirty-Three
Three Faces of Information

Scope: The 19th-century attribution of physical reality to patterns and relationships challenged the dominant view that the real was substantial. The idea that information structures, too, are physically real extends this challenge. Computers and the Internet have initiated a continuing explosion in content available "online"; this content explosion is real and affects communication, as well as the conduct of scientific research, commerce, and politics worldwide. But content is not the most fundamental sense in which we live in an information age. Claude Shannon's post–World War II, content-independent mathematical theory of information made information a feature of physical reality. This apparently abstract theory has become the foundation of powerful information technologies that continue to change the world. Concurrently, the idea that DNA encodes the "secret" of life in a base-sequence information structure and the idea that black holes and even the Universe itself may be information structures have reinforced the physical reality of information.

Outline

I. We may well live in an information age, but few people realize how many different meanings *information* has.

   A. The most familiar meaning of *information* is related to content.
      1. It is useful to distinguish information from data, on the one hand, and from knowledge, on the other.
      2. Although they are reciprocally correlated, information can be understood as organized data, while knowledge is interpreted information.
      3. This is not a hierarchical relation, because it is typically the case that data and their organization are informed by ideas, for example, ideas of what kinds of data and what forms of organization are likely to prove valuable.

   B. More information is available to more people than ever before, but this is not attributable only to the computer.

      1. The dissemination of more books to more people as the literacy rate increased was already a fact of 19th-century life.
      2. At the same time, mass-circulation newspapers and, by the end of the century, magazines created an information problem.
      3. By the late 20th century—with the telephone, radio, and television in addition to books, newspapers, and magazines—not even scholars could keep up with their disciplines.
  C. The Internet exacerbated an existing problem; it didn't create the problem.
      1. The idea that motivated the creation of the Internet was to give computer users direct, real-time access to files stored on other computers.
      2. The search engine Mosaic and the World Wide Web unintentionally accelerated the process of consumerizing and commercializing what had been a network oriented toward the computer community.
      3. This was reinforced by the creation of Netscape, Yahoo, AOL, and Google and the evolution of global interactive video games and virtual communities.

II. At the opposite pole from this everyday, familiar notion of information is a counterintuitive, mathematical conception of information that is the cornerstone of both computer and communication technologies.
  A. In 1948, Claude Shannon published an epochal paper, "The Mathematical Theory of Communication."
      1. Shannon was a mathematician at Bell Labs who had received degrees in electrical engineering and mathematics from MIT.
      2. As a graduate student, he had worked on, and been impressed by, Vannevar Bush's Differential Analyzer.
      3. His master degree thesis, "A Symbolic Analysis of Relay and Switching Circuits," explored the relationship between Boole's algebra of reasoning and digital telephone switching circuits.
      4. Aware of Shannon's evolving ideas about information, Bush urged him to apply them to Mendelian genetics, leading to his doctoral dissertation, "An Algebra for Theoretical Genetics."
      5. At Bell Labs during World War II, Shannon developed probability-theory–based tools for predictive antiaircraft fire control systems, modeling intelligence as signal processing.

6. Shannon's Ph.D. thesis brought him to the attention of Warren Weaver, with whom he co-authored *The Mathematical Theory of Communication* in 1948.

B. Shannon's theory strips information of meaning and connects it to thermodynamics.
1. *Information* is defined in terms of uncertainty, such that the greater the uncertainty, the greater the measure of information; information is treated as a statistical difference between what is known and what is unknown.
2. This implies that knowing the content of a message means the message is totally uninformative!
3. Shannon's problem was how an automated receiver could distinguish an incoming signal from random noise.
4. The content of the signal was irrelevant: All that mattered was recognizing the presence of a signal being transmitted and accurately reproducing it.
5. Communication becomes a stochastic process and semantics is dismissed as irrelevant.

C. Shannon showed that his problem divided into three quasi-independent problems.
1. The source problem was defined by the nature of the signal being transmitted.
2. The channel problem was defined in terms of how the signal was encoded at one end and decoded at the other.
3. The receiver problem reduced to the accuracy with which the decoded signal could be reproduced.
4. The information content of a message reduced to the number of bits required to transmit it accurately, regardless of what it was about.
5. To model this process, Shannon adapted mathematical techniques developed by MIT professor Norbert Wiener, founder of cybernetics.
6. Surprisingly, the equation correlating information and uncertainty is identical in form to the equation relating entropy to temperature in thermodynamics.
7. This finding suggested that mathematically, information and entropy are related.

- **D.** This highly technical, semantics-free interpretation of information has been astonishingly fertile in practical applications.
    1. Shannon's theory underlies all fiber optic technologies. Cell phone electronics and software are also based on Shannon's theory, as are recording and reproduction technologies for music CDs and video DVDs.
    2. Shannon argued that a digital/binary *bit* stream optimized encoding, decoding, and error detection and correction processes.
    3. Shannon noted early on that information storage is really a form of delayed communication; thus, his theory underlies generations of evolving information-storage technologies.
    4. Shannon's early work relating George Boole's laws of thought to telephone switching circuits led to designs for the logic circuits of digital computers that are built into the overwhelming majority of today's computer chips.
    5. In the 1950s and 1960s, at least, Shannon and a Bell Labs colleague were said to have applied his mathematical models to gambling and the stock market with considerable success.

III. The commonsense understanding of our age as an information age and a Shannon-based understanding of information overlap inside technologies that surround us, and they intersect a third sense of *information* that is different from both.
- **A.** Information—like fields, structure, and relationships—has been elevated to the status of an elementary feature of reality.
    1. Consider, for example, the discovery of the genetic code.
    2. In the 1930s, DNA was explicitly dismissed as the key to genetics, in spite of its universality, precisely because it was chemically the same in all life forms, always containing the same four bases in the same proportions.
    3. In the wake of the Watson-Crick discovery of the molecular structure of DNA, George Gamow suggested that the key to the functionality of DNA lay in a code: the precise sequence of the four bases that bind the strands of the double helix.
    4. It is not the bases that differentiate one life form from another but their sequence, which is, of course, an abstraction.
    5. The sequence is physically real without being material in the most fundamental sense of scientific reality: causal efficacy.

      6. The sequence defines the work that the DNA does in each cell of an organism, directing the production of proteins via RNA molecules, whose work is also defined by the parallel sequence of bases.

      7. Furthermore, the information structures that define each life form work through the formal structures of the proteins for which the sequence codes.

B. Information structures appear as elementary realties in astrophysics and cosmology, not just molecular biology.

      1. The theory of black holes uses information theory mathematics, which itself draws on the same mathematical rules that define entropy and thermodynamics.

      2. Moreover, black holes can be understood as structures that *preserve* information—a conception of black holes that Stephen Hawking, after decades of dismissing it, now accepts.

      3. This recognition has reinforced the *holographic principle*, which suggests that not only black holes but the entire Universe is an information structure.

      4. As extended and developed by Russian mathematician Andrey Kolmogorov as well as Americans including Gregory Chaitin and Ray Solomonoff, Shannon's information theory has evolved into *algorithmic information theory*, in which physical objects are "reduced" to information representations.

Essential Reading:

Charles Seife, *Decoding the Universe.*

Hans Christian von Baeyer, *Information: The New Language of Science.*

Questions to Consider:

1. Is information really just another name for knowledge? If so and if information is physically real, is knowledge also physically real?
2. In that case, what happens to the subjectivity-objectivity/mind-world distinctions?

Lecture Thirty-Four
Systems, Chaos, and Self-Organization

Scope: Atomistic thinking presumes that we discover the truth about reality by decomposing physical, chemical, biological, and social phenomena into elementary building blocks from whose properties we can synthesize the world. Three closely related 20th-century ideas challenge the adequacy and the correctness of this presumption: the ideas that phenomena are produced by systems, that apparently chaotic "real-world" systems are orderly, and that natural and social phenomena are produced by self-organizing systems. The distinctive features of these ideas are that systems display properties that are not displayed in, or derivable from, the individual properties of their constituents. In other words, that wholes are causally more than the sum of their parts; that the concept of lawfulness, already stretched by having to accommodate statistical "laws" and randomness, must include orderliness of a precise mathematical sort but without predictability; and that natural and social phenomena maintain themselves in stable, long-lived, non-equilibrium states via energy inputs.

Outline

I. In the two previous lectures, "computer" and "information" were single names, but each referred to three very different ideas; in this lecture, *systems*, *chaos*, and *self-organization* are three different names for a single complex idea.

    A. Systems theory, chaos (or complexity) theory, and self-organization theory are three facets of one idea.

        1. Both chaos theory and self-organization are characteristics of systems and, thus, presuppose the existence of a system.

        2. Only some systems are chaotic or self-organizing, but there is an intimate connection between chaos and self-organization and, of course, between both and the dynamics that distinguish systems from non-systems.

    B. Systems thinking contrasts sharply with atomistic, "bottom up" thinking.

1. Atomistic thinking assumes that entities with fixed, inherent properties are the elementary explanatory or compositional units of reality.
2. All experiential phenomena are the results of typically complex interactions among these entities.

C. Systems thinking is "top-down" thinking.
1. The systems approach is process-based rather than atomistic, and it emphasizes structure/form and relationships.
2. A system is a distinctive organization of mutually adapted parts, the nature of the adaptation being such as to enable the functionality of that specific system.
3. In a system, the parts cannot be analyzed independently of their position and function in the whole.
4. Systems have a holistic character: A system functions only as a dynamic whole of parts working together.
5. As a whole, a system displays *emergent* properties that its individual parts do not have: Typically, this is what is meant by "The whole is greater than the sum of its parts."

D. These emergent system-level properties are as real and as causal as the properties attributed to "atoms" in the atomistic approach to nature.
1. Molecules are, in this sense, mini-systems: Even a simple molecule, such as sodium chloride, has properties that neither of its atoms separately possesses while lacking properties that these atoms possess on their own.
2. The functioning of a complex molecular-system, such as DNA, powerfully illustrates the causal efficacy of emergent properties.
3. The same is true of a large-scale biological system, such as an ant colony, which also displays self-organization and reorganization, if disrupted, without a central controlling authority.

E. Nature is pervasively systemic.
1. Moreover, natural phenomena typically display a hierarchical system architecture based on coordinated modular subsystems.
2. Herbert Simon has argued that such a module-based hierarchical structure has strong evolutionary advantages.

3. The cell, for example, is a system with modular subsystems of its own, but the cell as a whole is also a modular subsystem within a grand hierarchy of tissue systems, organized into organ systems, organized into organisms. In turn, these organisms are "modules" within ecological systems that are themselves elements within a total environmental system that is the Earth.
4. Indeed, plate tectonic theory strongly reinforced the growing recognition, in the 1960s, that the Earth was a system, whose largest interacting units were the atmosphere, the oceans, the land, and the dynamic internal structure of core, mantle, and crust.

F. The properties of a system are determined by the form of its organization.
1. The relationships among parts and their mutual adaptation to one another and to the functionality of the whole are the determinative features of a system.
2. George Cuvier used this principle to reassemble the skeletons of fossil animals from their jumbled bones.
3. In artificial systems, these relationships and this adaptation are obviously the result of deliberate design decisions.
4. Because such adaptation is pervasive in nature, some people find the idea of intelligent design attractive, but the Darwin-Wallace theory of evolution proposes a process that generates apparent design spontaneously.

II. In 1963, one Earth subsystem, the atmosphere, gave rise to *chaos theory*.

A. Accurate weather prediction requires modeling the atmosphere as a system.
1. In 1963, meteorological physicist Edward Lorenz noticed the exquisite sensitivity of his then state-of-the-art atmospheric model to minute variations in initial conditions.
2. Following up on this observation led Lorenz to recognize that the atmosphere is a complex system that is composed of associated (weather) subsystems and is non-linear and "chaotic" in its behavior.
3. That is, weather systems interacting within the total atmospheric system are exquisitely sensitive to even small

variations in the parameters that determine them, such that small changes in these parameters at one place can lead to different weather conditions thousands of miles away.
4. Jokingly, this was called the *butterfly effect*, and it explains why long-term weather forecasting is so difficult.

B. The behavior of such non-linear systems is not predictable using the tools and assumptions of traditional Newtonian deductive-deterministic physics, nor is it periodic or equilibrium-seeking.
1. Lorenz and a growing community of mathematicians and scientists across disciplines began to study chaotic systems.
2. What emerged was a recognition that these systems were not chaotic at all, in the sense of hopelessly disordered and anarchic, which led to changing the name *chaos theory* to *complexity theory*.
3. Complex systems, characterized by an exquisite sensitivity to small variations in critical parameters, display a new kind of order, one that is well described by a family of mathematical equations.
4. Just as scientists in the 19th century had to redefine the lawfulness of nature to accommodate statistical laws, now they had to redefine order to accommodate system structures that were non-linear, non-periodic, and non-predictable in detail yet were stable, maintaining a distinctive structure over time far from equilibrium.
5. A hurricane is a dramatic example of such a structure, emerging out of atmospheric and oceanic interactions, growing and maintaining itself over weeks, and modifying its structure as it interacts with other systems.

C. The motivation for this redefinition was the wide range of applications of these equations to natural and social phenomena.
1. The study of the mathematical equations that described "chaotic" behavior turned out to be applicable to a much wider range of phenomena than weather systems.
2. Nature is pervaded by non-linear systems, but before the 20th century, modern science simplified these to linear systems because that's what the available mathematics could describe.
3. Social phenomena, for example, the fluctuations of the stock market, also lend themselves to analysis by these equations.

    4. But weather systems also display another underappreciated behavior: They are self-organizing, as in the cited case of the hurricane.

III. Self-organization theory is a subset of general systems theory and closely connected to chaos/complexity theory.
  A. In the 1970s and 1980s, chemist Ilya Prigogine emerged as the champion of self-organization theory.
    1. Initially, Prigogine called attention to and studied fairly simple chemical systems that self-organized into fairly complex structures that were stable over time.
    2. In effect, all you need to do is combine the right ingredients, "mix," and stand back: Structures emerge by themselves.
    3. Prigogine's research expanded to include recognition of self-organization as a general feature in nature and especially of chaotic/complex systems.
    4. What they all have in common is that the structures that emerge are non-linear, stable far from equilibrium, and adaptive: Within limits, they can evolve in response to environmental change.
    5. An extraordinary example of self-organization is the development of the human embryo.
  B. The spontaneous emergence of order in nature challenges the idea of entropy in thermodynamics.
    1. The second law of thermodynamics implies that entropy in a closed system must increase over time and orderliness must decrease.
    2. But with an available source of energy, systems can self-organize out of available, unordered ingredients and spontaneously generate stable structures that can themselves evolve over time into more complex structures.
    3. In fact, just this property is central to emerging nanotechnologies and is already being exploited commercially.
    4. John Holland and Arthur Samuel, around 1950, began to explore the spontaneous emergence of order in the context of computer programs that could learn from their own experience.

5. Samuel created a program that became the world checkers-playing champion, and Holland created a family of programs called *genetic algorithms*.

IV. All this is a profoundly new mindset for scientists, one that has become interwoven with theories of science as well as with our expectations about how we use technology.
- **A.** We are much more sensitive today to the complexity of the Earth as an environmental system.
- **B.** The most marvelous expression of self-organization is the formation of the human embryo, which spontaneously emerges out of non-linear interactions.
- **C.** Hierarchical systems structures are characteristic of almost all natural phenomena and also of technological systems fundamental to how we live our lives—such as the telephone system, the electrical grid network, and the Internet.

Essential Reading

John H. Holland, *Hidden Order: How Adaptation Builds Complexity*.

Ilya Prigogine, *Order Out of Chaos*.

Questions to Consider:

1. If systems themselves emerge out of self-organizing preexisting elements, then isn't atomistic substance metaphysics right after all?
2. Does the centuries-long prejudice favoring equilibrium as the norm suggest that our thinking is always shaped by analogous prejudices?

Lecture Thirty-Five
Life as Molecules in Action

Scope: The explanation of all life phenomena (except for the human mind) in terms of the deterministic motions of lifeless matter was central to Descartes' mechanical philosophy of nature. The rival, vitalist, view that life was not reducible to lifeless matter came under increasing pressure in the 19th century, especially from biophysics- and biochemistry-based research programs. From a focus on the organism and its behaviors, biology shifted to a focus on the cell and its molecular processes, finding expression in the enzyme/protein theory of life, reinforced by the discovery that proteins were configurations of amino acids. Sealing this shift was the further discovery that DNA molecules—and within each of them, a distinctive sequence selected from the same four bases—defined every life form on Earth. By the 1980s, the molecular theory of life was transforming medicine by way of a flourishing biotechnology industry based on its research findings and transforming the meaning of life, as well.

Outline

I. The claim that life is a matter of specific molecular processes, and no more than that, is both old and new.

   A. A radically materialistic interpretation of life goes back to Greek antiquity and was built into modern science.

      1. Epicurus's atomic theory of matter, based on the earlier ideas of Democritus and Leucippus, was itself the basis of a rigorously materialistic theory of life, mind, and soul.

      2. Epicurus's ideas were disseminated in the Roman world by Lucretius and, through him, to the Renaissance.

      3. Descartes' mechanical philosophy of nature claimed that all natural phenomena, including all life phenomena, were the result of matter in motion.

      4. For Descartes, animals were machines (*bêtes machines*), which justified vivisection to study these complex devices.

5. Humans were machines, too, in their bodily subsystems, but not the mind, which Descartes excepted as an immaterial entity.
6. Eighteenth-century Cartesians defended an exhaustive materialism that included the mind, stimulating the first attempts at reducing mind to brain.

B. In the 19th century, the idea that life and mind were strictly physico-chemical phenomena rapidly grew to dominance.
1. With respect to the mind, advances in neurophysiology; the beginnings of experimental psychology; and the work of Pavlov, John Watson, and even Freud brought this topic within the domain of mechanistic science.
2. With respect to the body, the cell theory of life was developed largely by chemists and biologists committed to a mechanical interpretation of life.
3. Indeed, by the late 19th century, basic research in biology was increasingly a matter of biochemistry, i.e., of understanding life phenomena by studying cell structures and reactions in the cell.
4. Emil Fischer's discovery that proteins are built up out of amino acids and can be synthesized in the laboratory strongly reinforced the mechanistic view.

II. At the same time, molecular biology is a new idea.
A. The term seems to have been used first in 1938 by Warren Weaver of the Rockefeller Foundation, responsible for research grants in the biological sciences.
1. Weaver was committed to strengthening physics and chemistry as the basis of truly scientific biological research: To understand biological phenomena meant "reducing" them to chemical and physical phenomena.
2. This mindset in biology became pervasive, however, only in the wake of Watson and Crick's discovery of the molecular structure of DNA in 1953.
3. Initially at least, DNA was considered to contain all by itself the secret of life and to function as an absolute monarch dictating all life processes.

B. But the wholesale embrace, after 1953, of molecular biology as the foundation of all biological research clearly built on other shifts within science that began in the late 19th century.
   1. The rise of quantum theory, in principle, reduced chemistry to physics.
      a. Chemical reactions were believed to be based on exchanges of orbital electrons.
      b. Into the mid-1930s, with the rise of nuclear physics, quantum mechanics was a theory of orbital electrons.
      c. Although the quantum mechanical equations for real chemical reactions were too difficult to solve, in principle, chemistry was physics and biochemistry was biophysics.
   2. The Cartesian claim that even life was just matter in motion received a much more sophisticated formulation.

C. One expression of the impact of quantum physics on chemistry and biology was the spread of new physics and physical chemistry instruments and, later, of physicists themselves into the biology laboratory.
   1. X-ray crystallography is a preeminent example: it was based on 20th-century physics and initially developed to test physics theories.
   2. X-ray crystallography is immediately applicable to determining the structure of any molecule that can be crystallized without disrupting the arrangement of its atoms.
   3. Obviously, this applied just as well to the structure of molecules of interest to biologists.
   4. Concurrently, Swedish physicist Theo Svedberg developed the ultracentrifuge to settle an important issue in chemistry: whether very large molecules were robust or loose conglomerations of much smaller molecules.
   5. Svedberg's machine quickly showed that Hermann Staudinger's defense of macromolecules was correct, though Hermann Mark showed that these molecules were flexible, not rigid.

D. Mark used these new instruments intensively in his research and became an influential founder of polymer chemistry.

1. Linus Pauling, one of the earliest chemists to apply quantum mechanics to the analysis of chemical bonds, visited Mark's lab in 1930.
2. Pauling was impressed with Mark's experimental methodology and, on his return to the United States, aggressively pursued funding to buy new instruments for his lab.
3. He succeeded eventually in receiving sustained Rockefeller Foundation funding through Warren Weaver, beginning around 1937.
4. By that time, the electron microscope had been invented and was being produced commercially.
5. Weaver was already committed to the ultracentrifuge, supporting Svedberg's research in Sweden in exchange for providing the machine and ongoing updates to the Princeton University biology department.

E. In the 1930s and 1940s, Rockefeller Foundation money "pushed" the latest physics-based instruments keyed to biochemical research into many U.S. biology labs.
1. The results were impressive, including Dorothy Crowfoot Hodgkin's discovery of the molecular structure of penicillin and vitamin B12.
2. Around 1940, Linus Pauling demonstrated that the antigen-antibody reaction was a matter of molecular structural properties and, later, that sickle cell anemia was a structural flaw in blood cells caused by a misplaced amino acid in one protein.
3. In 1951, Pauling showed that RNA had an alpha-helical structure, but he lost the "race" to discover the structure of DNA to Watson and Crick, who used the X-ray crystallographic data collected by Rosalind Franklin and her collaborators.

III. The excitement surrounding the discovery of the structure of DNA made molecular biology the hallmark of state-of-the-art biological research.
A. By 1957, with Matthew Meselson and Franklin Stahl's demonstration that DNA replicates by "unzipping" and forming new complementary helices, DNA was hailed as the key to understanding life.

1. Understand how DNA determines the chemical reactions that make up all cellular processes and you understand life.
2. In 1912, Jacques Loeb had written an influential book titled *The Mechanistic Conception of Life*, but when Paul Berg and his collaborators announced the success of their recombinant-DNA experiments in 1972, molecular biology became practically the only game in town.

B. In the course of the next two decades, molecular biology of DNA became the cornerstone of the commercial biotechnology industry.
1. Biotechnology based on DNA manipulation has already had a profound influence on clinical medicine and promises vastly greater influence in the 21st century.
2. By 1982, the FDA had approved the sale for humans of insulin "manufactured" by genetically modified bacteria, and the list of similar substances continues to expand.
3. In 1980, the first transgenic plants and animals were created, and the former especially has led to the steadily growing genetically modified food and agricultural industry.
4. Transgenic animals are having a major impact on research into human illnesses and the effectiveness of new drugs.
5. Stem-cell research promises extraordinary new possibilities, but whether these will be fulfilled is a 21st-century story.

Essential Reading

Michel Morange, *A History of Molecular Biology*; *The Misunderstood Gene*.

James B. Watson, *The Double Helix*.

Questions to Consider:

1. What are the implications for our belief that life is special in the growing power of molecular biology?
2. If genes act only in conjunction with factors internal and external to the organism, how could a human clone be more than superficially similar to a donor?

Lecture Thirty-Six
Great Ideas, Past and Future

Scope: Scientific ideas, built on the ideas of science and techno-science, have changed the world and, for the foreseeable future at least, will continue to do so. At the "highest," most abstract level, the historic dominance of Parmenidean substance metaphysics has given way to a progressively greater role for Heraclitean process metaphysics. The idea that all natural phenomena were produced by chemical "elements" made up of truly elementary particles has been qualified by the idea that also elementary are hierarchically ordered systems manifesting relational and information structures and self-organization to sustain lawful but unpredictable, far-from-equilibrium states. Self-organization is fundamental to the emerging nanotechnology industry, which promises to become the defining technology of the early 21st century, while a synthesis of Parmenidean and Heraclitean metaphysics is fundamental to string theory and its highly controversial attempt to unify the forces of nature into a comprehensive theory of everything.

Outline

I. The mosaic "big picture" that emerges from the individual lectures in this course is that science, allied to technological innovation, became the dominant driver of social change because society empowered it to play that role.

   A. The biggest single piece of that mosaic picture was the emergence in ancient Greece of the idea of science itself.
   1. That ancient idea of science erupted in the 17th century as modern science.
   2. It is an expression of particular definitions of *knowledge*, *reason*, *truth*, and *reality* and the relations among them.
   3. The invention of mathematics as a body of knowledge matching these definitions and the connection of mathematics to knowledge of nature were fundamental to the idea of science.

   B. One lesson to be learned from this Greek invention of the idea of science is the centrality of definition to science.

1. Scientific reasoning inevitably begins with definitions of terms and concepts, but these definitions are contingent.
2. Scientific definitions are not "natural"; they must be proposed, and they change over time, as scientists' explanatory objectives change.

C. The idea of techno-science also emerged in antiquity, in the Graeco-Roman period.
1. Although the idea of techno-science may seem an obvious extension of the pursuit of knowledge of nature, it is not.
2. Plato and Aristotle explicitly excluded know-how from knowledge and explicitly separated understanding from action.
3. This elitist view of theory-based science as different from, and superior to, practice-based technology was prominent in the 19th century and continues today.
4. Secondly, Graeco-Roman science was so modest that the idea that science could be the basis of powerful new technologies was truly visionary.

II. The second largest piece of the mosaic was the emergence of modern science in the 17th century.
A. The second cluster of lectures aimed at showing that the rise of modern science was an evolutionary development, not revolutionary.
1. One of the roots of modern science is the rebirth of cultural dynamism in western and central Europe in the period 1100–1350.
2. The invention of the university recovered and disseminated to tens of thousands of students the Greek ideas of knowledge, logic, mathematics, and knowledge of nature.
3. The university also disseminated the secular and naturalistic values that are implicit in pursuing knowledge for its own sake.
4. These values were reinforced by the concurrent industrial revolution, which recovered and innovated upon Graeco-Roman know-how in agriculture and industry in conjunction with expanded commerce.
5. This created a broad and deep body of mechanical know-how in Western society.

B. A lesson about the science-society relationship from this cluster of lectures is the dependence of the social impact of science and technology on society.
   1. Technological innovations, as Lynn White said, merely open doors.
   2. The same innovations have had very different impacts on different societies.
   3. In each case, the way innovations are implemented in a society reflects values and institutions characteristic of that society, not of the technology.
   4. But this is just as true of scientific knowledge as of technological know-how: Ideas are analogues of innovations.

C. A second root of modern science is the culture of the Renaissance (1350–1600).
   1. In the 15th and 16th centuries, the idea of progress was explicitly identified with growth of knowledge and knowledge-based know-how, techno-science.
   2. The Humanists made important Greek mathematical and scientific texts available to the Latin-speaking West, and they developed critical methodologies for recovering corrupted texts that influenced the scientific method.
   3. Mathematics became the basis for a wide range of new technologies, reinforcing the view that mathematics was the key to knowledge of nature.
   4. The response of Western societies to Gutenberg's new print technology illustrates how technologies merely open doors and created the medium modern science would adopt.

D. The emergence of modern science reflects its indebtedness to antiquity, the late Middle Ages, and the Renaissance.
   1. The ideas of knowledge, deductive reasoning, mathematical physics, and progress were all part of the intellectual environment, as was the university as an institution that taught these ideas at an advanced level.
   2. The accomplishments of the founders of modern science in the 17th century are built on these ideas, including the idea that knowledge of nature will give power over nature and that this is the key to progress.

III. Modern science matured in the 19th century, and its theories and ideas became fertile from the perspectives of understanding nature and acting on nature.
- **A.** These theories reflect two distinct approaches to the study of nature: the substance or atomistic approach and the process approach.
    1. The atomic theory of matter, the cell theory of life, the germ theory of disease, and the gene theory of inheritance, in their initial formulations, all reflect the atomistic approach.
    2. Theories in physics and chemistry based on new ideas of energy, fields, and structure reflect the process approach.
    3. The Darwin-Wallace theory of evolution explains the range of living things as the changing outcome over time of an ongoing process.
    4. Concurrently, the idea of statistical laws underlying natural and social phenomena challenged the determinism implicit in the idea of knowledge as scientists had assimilated it.
    5. Quantum mechanics is our deepest theory of matter and energy, and it is a probabilistic theory.
- **B.** Twentieth-century theories reveal a continuing challenge to elementary substance approaches to nature.
    1. The context of 20th-century scientific theories reveals the influence of new institutions created specifically to promote science-based technological innovation.
    2. This reinforces the extension of Lynn White's dictum about the relationship of technology to science: The science-society connection is dynamic and reciprocal, such that neither is independent of the other.
    3. Twentieth-century science and technology reflect the increasingly cross-disciplinary character of scientific research and its applications, which couples society back to science through the response of society to technological innovations.

IV. We can, cautiously, anticipate scientific ideas and technological innovations that will transform 21st-century life.
- **A.** The social impact of technological innovations is unpredictable, but at least two families of innovations can be predicted with confidence to open many new doors in the coming decades.

1. We are just beginning to develop commercial applications of nanotechnologies, but these will almost certainly define a new stage of techno-science.
2. It is already clear that self-organization is the key to the successful mass production of nanotechnological devices, and this is another illustration of how an abstract idea or theory can have important practical applications.
3. The second family of innovations is the continuing development of biotechnologies, which are inherently self-organizing.

B. There are two areas of science that can also confidently be predicted to have major social impacts in the next half-century.
1. One of these areas is the continuing naturalization of consciousness via molecular and computational neuroscience, with profound consequences for what it means to be human.
2. It is likely that new technologies for controlling moods, behavior, memory, and learning will be developed, posing deep ethical and moral challenges.
3. The other area is the pursuit by physicists of a theory of everything, a theory that unites the four fundamental forces of nature into a single, comprehensive, universal theory of matter and energy, indeed, of the Universe.
4. Quantum theory seems to be the best foundation for such a theory, given that the standard model already unites three of the four forces. String theory has the greatest visibility as the approach to full unification, but it has serious problems.

C. The pursuit of unification is interesting in and of itself.
1. The quest for unification seems perennial in Western culture.
2. Monotheism is the product of one unification, and Greek materialistic monism, another.
3. It is also interesting that so many modern scientific ideas echo ideas articulated by Greek philosophers more than 2000 years ago.
4. With some scientists claiming that we are now, *really*, on the threshold of unification, it is important to recall that we have been there before!

Essential Reading

Mark Ratner and Daniel Ratner, *Nanotechnology: A Gentle Introduction to the Next Big Idea.*

Leonard Susskind, *The Cosmic Landscape: String Theory and the Illusion of Intelligent Design.*

Questions to Consider:

1. Are we today better equipped to manage the social impact of emerging nano-, bio-, and mind-technologies than we were when managing the last generation of innovations?
2. If we are approaching the end of creative scientific theorizing, what are the cultural implications of such a development?

Timeline

9000 B.C.E.	Domestication of grains and fruits begins.
7000 B.C.E.	First evidence of copper smelting; evidence of drilled teeth.
6000 B.C.E.	Earliest evidence of wine making; large-scale settlements in the Middle East.
4500 B.C.E.	Horizontal loom weaving.
4000 B.C.E.	Modern wooly sheep.
3500 B.C.E.	Sumerian cuneiform writing.
3000 B.C.E.	Beer brewing in Sumer; earliest gold jewelry; twill weaving using warp-weighted loom.
2800 B.C.E.	Bronze in use in Sumer; Egyptian hieroglyphic writing.
2000 B.C.E.	Spoked-wheel chariots.
1800 B.C.E.	Egyptian medical and mathematical papyri; Babylonian Code of Hammurabi; alphabetic writing in Ugarit.
1500 B.C.E.	Iron manufacturing; cast bronzes in China; vertical loom weaving.
1300 B.C.E.	Earliest Chinese inscriptions; Phoenician alphabetic writing.
1250 B.C.E.	Glass manufacture in Egypt.
1000 B.C.E.	Steel making on a limited scale.
800 B.C.E.	Hellenic Greeks adopt Phoenician alphabet.
700 B.C.E.	Homeric epics written down.

500 B.C.E.	Cast iron in use in China.
5th century B.C.E.	Thales, Pythagoras, Parmenides, Heraclitus, Anaxagoras, Empedocles.
4th century B.C.E.	Plato, Aristotle, Euclid, Epicurus.
3rd century B.C.E.	Roman conquest of Greece; Archimedes, Apollonius, Aristarchus.
2nd century B.C.E.	Antikythera machine.
1st century B.C.E.	Vitruvius; Chinese invention of paper.
1st century C.E.	Pompeii buried by Vesuvian ash; Frontinus on aqueducts of Rome.
2nd century C.E.	Hero of Alexandria's book of machines; Baths of Caracalla in Rome; watermill complex near Arles; Ptolemy and Galen.
451	Conquest of Rome by Goths.
521	Justinian closes Athenian philosophy schools.
1086	Domesday Book inventory of England.
1092	Start of the First Crusade.
1170	Universities of Bologna and Paris founded.
1268	First weight-driven mechanical clock.
1329	Start of the Hundred Years' War between England and France.
1347	First outbreak of plague in Europe.
1350	Petrarch founds the Humanist movement/idea of progress.
1415	Brunelleschi rediscovers perspective drawing.

1425	Jan van Eyck introduces oil-based paints.
1453	Gutenberg introduces printing with movable metal type.
1487	Vasco da Gama sails around Africa to India.
1492	Columbus's first voyage to the New World.
1512	Michelangelo completes the Sistine Chapel ceiling painting.
1515	Ferdinand Magellan begins first around-the-world voyage.
1543	Copernicus's *On the Revolutions of the Heavenly Spheres*; Vesalius's *On the Structure of the Human Body*.
1554	Gerard Mercator's mathematics-based maps of Europe.
1600	William Gilbert's *On the Magnet*; Giordano Bruno burned at the stake in Rome.
1609	Kepler's *New Astronomy* claims elliptical planetary orbits.
1610	Galileo's telescope-based *Sidereal Messenger*.
1618	Start of the Thirty Years' War.
1619	Kepler's *Harmony of the World*.
1620	Francis Bacon's *New Organon*.
1628	William Harvey's *On the Motion of the Heart and Blood in Animals*.
1632	Galileo's *Dialogue Concerning the Two Chief World Systems*.

1637	René Descartes introduces algebraic geometry.
1638	Galileo's *Discourses on Two New Sciences*.
1648	Treaty of Westphalia ends the Thirty Years' War.
1660	Royal Society of London founded.
1665	Robert Hooke's *Micrographia*.
1666	French Royal Academy of Science founded.
1673	Anton Leeuwenhoek's first published microscope observations.
1684	Leibniz's first calculus publication.
1687	Newton's *Principia Mathematica*.
1704	Newton's *Opticks*; Newton's first calculus publication.
1709	Jacob Bernoulli introduces modern probability theory.
1750	Thomas Wright's Newtonian cosmology.
1758	John Dollond patents color-corrected microscope lens.
1767	James Hargreaves's spinning jenny.
1771	Richard Arkwright's water-powered spinning "frame."
1776	U.S. Declaration of Independence; Adam Smith's *The Wealth of Nations*.
1776	Watt-Boulton steam engines commercially available.
1782	Lavoisier discovers oxygen, initiates chemical revolution.

1789	French Revolution.
1794	Erasmus Darwin's poem *The Botanic Garden*.
1799	Laplace's *Celestial Mechanics*.
1800	Volta invents the electric battery.
1807	John Dalton introduces modern atomic theory; 1807.
1807	Georg Friedrich Hegel's *Phenomenology of the Spirit*.
1807	Robert Fulton's *Clermont* steamboat.
1809	Jean-Baptiste Lamarck's *Zoological Philosophy*.
1822	Joseph Fourier's analytical theory of heat published.
1824	George Boole's laws of thought; Sadi Carnot's *Reflections on the Motive Power of Heat*.
1828	George Stephenson's *Rocket* steam locomotive.
1830	Michael Faraday invents the dynamo.
1835	Charles Darwin returns from his global voyage on H.M.S. *Beagle*.
1835	Adolphe Quetelet founds social statistics.
1838	Friedrich Bessel measures the distance to star Cygnus 61.
1838/39	Mathias Schleiden and Theodore Schwann's cell theory of life.
1844	Samuel F. B. Morse's pilot installation of an electric telegraph.
1845	Faraday introduces the field concept.

1847	Hermann Helmholtz proclaims conservation of *Kraft* (meaning "force" or "power"); the term "energy" would be introduced in 1850.
1849	Louis Pasteur discovers two forms of tartaric acid crystals.
1850	William Rankine coins *energy* for *Kraft*; Rudolph Clausius founds thermodynamics, coins the term *entropy*.
1851	William Thomson proclaims the arrow of time.
1856	William Perkin discovers the first synthetic dye.
1857	Pasteur's essay on fermentation founds the germ theory of disease.
1858	Alfred Russel Wallace's essay on evolution by natural selection.
1859	Darwin's *On the Origin of Species*.
1862	Morrill Land Grant Act triggers growth of engineering education in the U.S.
1865	Mendel publishes results of his researches.
1865	Maxwell's *A Dynamical Theory of the Electromagnetic Field*.
1865	Auguste Kekule announces the ring structure of the benzene molecule.
1865	First effective transatlantic telegraph cable.
1877	Robert Koch isolates the cause of anthrax.
1879	Pasteur introduces modern vaccination.

1882	Koch isolates tuberculosis bacterium and, a year later, cholera.
1882	Thomas Edison inaugurates centrally generated electricity.
1885	Pasteur shows that dead bacteria confer immunity.
1895	Roentgen discovers X-rays.
1896	Henri Becquerel discovers radioactivity.
1898	Marie Curie names *radioactivity*, isolates polonium, then radium.
1897	J. J. Thompson discovers the electron.
1900	Hugo de Vries and others rediscover Mendel's results; Max Planck's quantum hypothesis.
1903	De Vries's *The Mutation Theory*; William Bateson coins the term *genetics*.
1903	Ernest Rutherford and Frederick Soddy determine lawful randomness of radioactive decay.
1905	Einstein's "miracle year" of publication.
1910	Thomas Hunt Morgan localizes the "gene" for fruit-fly eye color.
1910	Ernest Rutherford's Solar System model of the atom.
1912	Henrietta Leavitt Swann's variable-star cosmic "ruler."
1913	Niels Bohr's quantum theory.
1914	World War I begins.

1915	Einstein's general theory of relativity.
1918	World War I ends.
1923	Edwin Hubble announces that Andromeda is a galaxy.
1925	Heisenberg and Schrödinger found quantum mechanics.
1926	Heisenberg uncertainty principle; statistical interpretation of quantum mechanics.
1929	Hubble announces the expanding Universe; Paul Dirac founds quantum electrodynamics.
1931	Electron microscope invented.
1935	Karl Jansky detects radio signals from the Sun.
1935	First virus "seen" using electron microscope.
1936	Alan Turing publishes his principle of the universal computer.
1938	Warren Weaver coins the term *molecular biology*.
1939	World War II begins.
1945	First atomic bombs; World War II ends; ENIAC becomes operational.
1947	Transistor invented at Bell Labs.
1947–1949	George Gamow and colleagues propose the Big Bang theory.
1948	John von Neumann constructs EDVAC; Claude Shannon founds mathematical information theory.

1953	Watson and Crick announce the double-helix structure of DNA.
1956	Dartmouth conference on artificial intelligence.
1957	*Sputnik I* orbits the Earth.
1958	Jack Kilby and Robert Noyce invent the integrated circuit.
1963	Penzias and Wilson detect microwave background radiation; Edward Lorenz triggers chaos theory.
1964	Murray Gell-Mann and George Zweig found quantum chromodynamics.
1969	Neil Armstrong walks on the Moon.
1971	First test of ARPANet, leading to the Internet in the 1990s.
1971	Electro-weak unification wins acceptance.
1972	Recombinant DNA research begins.
1973	DEC introduces first "mini-computer" PDP-8.
1973	Standard model of quantum field theory formulated.
1980	Alan Guth's inflationary theory of the Universe; dark matter proposed.
1981	IBM PC introduced.
1984	String theory becomes "respectable" in physics.
1989	Disintegration of the Soviet Union.
1990	Hubble Space Telescope launched into orbit.

1991	Tim Berners-Lee introduces the World Wide Web.
1995	Sixth quark, called *top*, discovered.
1998	Dark energy proposed.
2000	Human genome decoded.

Glossary

aether: In 19th-century physics, a name for a universal space-filling form of matter or energy that served as the medium for immaterial fields of force.

algorithm: A series of well-defined operations that, if followed precisely, is guaranteed to solve a specified problem.

alphabetic: A name for a writing system that constructs the words of a language out of intrinsically meaningless symbols, in contrast with syllabic or ideographic writing systems.

amino acid: Typically in biology, this refers to one of 20 variants of a complex molecule—in which an NH_2 grouping of atoms shares a carbon atom with a so-called carboxyl group—out of which living cells construct proteins.

analytic geometry: Using algebraic equations to represent geometric forms and to solve problems in geometry or problems in algebra that previously had been solved using geometry, a Greek preference that was still common through the 17th century.

axiom: A statement whose truth is self-evident, hence, can be used for purposes of inference without requiring a proof that it is itself true. Sometimes used loosely for a statement that we are to take as true without further proof because true statements can be inferred from it deductively.

binary system: In arithmetic, a number system employing only two symbols, 0 and 1, to represent all possible numbers and the results of all arithmetic operations; more generally, any strictly two-place characterization of phenomena: for example, on-off or true-false.

block printing: Printing texts by carving the writing symbols into a block of wood that is then inked and pressed onto the writing surface.

cartography: Mapmaking.

central-vanishing-point perspective: A technique for creating the illusion of depth on a two-dimensional surface, for example, a wall or a canvas, such that the viewer "sees" the content of a painting as if the depth were real.

chromatic aberration: Distortion that worsens as magnification increases, caused by the fact that the focal point of a lens is a function of the frequency of the light rays striking it, while natural light contains multiple frequencies, leading to multiple foci and blurry images.

classification problem: Identifying classification categories that organize some set of objects into groups in ways that reflect features of those objects rather than values projected by the classifier.

coal tar: The residue, long considered waste, from burning coal in a closed vessel in order to generate a flammable illuminating gas that was sold for lighting (gaslight) and for cooking and heating.

coke: Burning piles of coal in a controlled way, such that only the outer layer of the pile is consumed, converts the inner material into coke, which can substitute for charcoal in iron-making, dramatically lowering costs.

concrete: A construction material composed of a binder or cement; an aggregate, typically sand and/or gravel; and water. Modern concrete, so-called Portland cement, is very similar to Roman cement, which used quicklime, pozzolana (volcanic ash soil), gravel, and water. The strength of concrete comes from the chemical combination of the water with the binder.

contingent: Dependent, for example, on context, time, or circumstances, hence, not necessary.

conventionalism: The view that classification categories—*tree, fish, planet*—have no reality apart from the individual objects being classified and reflect features of those objects that have attracted the selective attention of the classifier.

cosmos: Ancient Greek name for the Universe as an ordered whole, though not all Greek philosophers meant by *cosmos* everything that is, as we do.

cuneiform: A way of writing in which the symbols of a writing system are inscribed into a medium, for example, into clay tablets using a stylus.

deduction: A form of reasoning in which the truth of an inferred statement, the conclusion of a logical argument, is guaranteed, that is, follows necessarily, from the truth of some other statements, the premises of that argument.

demonstration: Literally, a "showing," but in reasoning, a deductive logical argument.

dialectic: In Greek logic, a form of reasoning in which the premises of an argument are not known to be true—either self-evidently or deductively—but are assumed to be true in order to explore their logical consequences.

digital: In modern technology, the representation of any phenomenon in discrete numerical terms, typically binary, in contrast with continuous analog representations. Where analog computers are customized to specific problems, digital computers have a universal character (if the numerical representation is valid!).

double-entry bookkeeping: Introduced to the West from Islam circa 1200 and popularized during the Renaissance, this method of record-keeping allowed precise tracking of debits and credits at a time when the scale and complexity of commerce were growing.

dynamo: A machine, commercialized from the 1870s, based on Michael Faraday's dynamo principle of about 1830, in which an electric current flows in a conductor as the result of the relative mechanical motion of the conductor and a magnet.

electrochemistry: Using electric currents to dissociate chemical compounds into their constituent elements, thereby identifying the constituents and permitting the study of "pure" elements.

electro-weak theory: The 1960s theory developed by Sheldon Glashow, Abdus Salam, and Steven Weinberg that unified the electromagnetic force, exerted by photons, and the so-called weak nuclear force, exerted by a family of particles called intermediate vector bosons, that is associated with radioactivity but also affects the electron family of particles, quarks, and neutrinos.

emergent property: A property of a whole that is not displayed by the individual parts of the whole *and* has causal consequences of its own.

empiricism: The view that all knowledge claims are ultimately validated by observational experience.

engineering drawing: A means of exhaustively characterizing the form of a three-dimensional object, however complex, using two-dimensional drawings. Renaissance engineers adapted perspective drawing to achieve this end and invented cutaway and exploded representations to show how machines were built; later, engineers and architects adopted orthogonal projections as a standard.

Enlightenment: A name given by 18th-century intellectuals to their age as one in which reason was used to improve mankind's physical, moral, social, and political condition.

enzyme: Enzymes are proteins that function as catalysts in metabolic processes; that is, they enable reactions but are not consumed in those reactions. Like all proteins, they are composed of amino acid complexes.

epicycles: A name for hypothetical centers of uniform circular motion for the planets to account for the fact that, viewed from the Earth, the planetary motions seem to be neither uniform nor circular.

ether: See **aether**.

feedback: A term popularized by Norbert Wiener's cybernetics theory in the 1940s, feedback refers to returning a portion of the output of a system or process to its input. Positive feedback reinforces the input, leading to a continually increasing output up to the limits of a system; negative feedback reduces the input and can, thus, be used to regulate the ratio of output to input.

floating point arithmetic: A scheme for representing numbers of any size compactly for purposes of automated calculation, as in a computer.

fractal: A name coined by the mathematician Benoit Mandelbrot to describe a family of shapes that occur throughout nature and are describable by a particular family of highly abstract mathematical functions. These shapes violate the traditional conceptualization of objects as being one, two, or three dimensional.

germ plasm: In the 1880s, August Weismann argued that sexually reproducing organisms inherit their distinctive character through a line of sexual cells, the germ plasm or germinal plasm, that is wholly isolated from the life experiences of the organism—hence, no inheritance of acquired characteristics, contra Lamarck. For us, DNA is the germ plasm.

hieroglyphics: An ideographic writing system, such as the one initially used by the ancient Egyptians in the 3rd millennium B.C.E. It evolved over the next 2000 years into increasingly stylized symbols that eventually represented syllables rather than ideas.

hydrostatics: The study of floating bodies in equilibrium, whose principles were first formulated by Archimedes in the 3rd century B.C.E. He also studied hydrostatics, the behavior of fluids, for example, water, in motion and the pressures they exert, which is directly relevant to the design and construction of water clocks and water- and air-powered machinery.

ideographic: A writing system in which each symbol, typically pictorial, expresses an idea.

induction: A form of reasoning in which the truth of some statement follows only with some probability from the truth of other statements; hence, it may be false even if those other statements are true, in contrast with deductive reasoning.

innovation: Not a synonym for invention but the form in which an invention is realized in the course of a process that integrates invention, engineering, and entrepreneurship.

kinetic theory of gases: The mid-19th-century theory developed especially by Clausius, Boltzmann, and Maxwell that the observable properties of gases—pressure, temperature, and viscosity—and relations among them are the result of, and are explained by, the motions of vast numbers of unobservable atoms or molecules of which they are composed; furthermore, these motions have only a statistical description.

logic: The name given to the study of forms of reasoning and their rules, independent of what the reasoning is about.

logical proof: See **proof**.

metaphysics: The study of the ultimately real, as opposed to what only appears to be real. The term occurs first as a description of an otherwise unnamed text of Aristotle's that deals with the first principles of being.

modern science: A name for an approach to the study of natural phenomena that emerged in the 17th century, was extended to social phenomena in the 18th century, and is considered to have developed into what we mean by *science* today.

mutation: A discontinuous variation typically in some heritable attribute of a cell, organism, or today, a DNA molecule, by comparison with its "parent." Hugo de Vries made mutations the basis of his 1901 theory of inheritance and of evolution, replacing natural selection with mutations.

nanotechnology: The manipulation of matter and the creation of structures on a molecular scale, with features measured in nanometers, billionths of a meter, or about 10 angstroms on an alternative scale. A DNA molecule is about 4 nanometers wide, and the read/write head of a state-of-the-art hard drive floats about 3 nanometers above the disk.

naturalism: The view that nature is the sum total of what is real, a view espoused by Aristotle against Plato's view that the ultimately real were supra-natural forms.

nucleosynthesis: The synthesis of the rest of the elements out of the hydrogen that is assumed to have been the universal form of matter in the early Universe.

perspective drawing: See **central-vanishing-point perspective**.

plasm: See **germ plasm**.

polymer chemistry: The study and manipulation of the properties of large molecules built on long, linear chains of carbon atoms. Plastics are polymers.

population genetics: Statistical models of the distribution of genes in large populations of randomly breeding individuals. Conceptually, the development of population genetics in the 1920s echoes the ideas underlying the kinetic theory of gases, statistical mechanics, and thermodynamics and the statistical laws of radioactivity and quantum mechanics.

process metaphysics: The view that reality is ultimately characterized by, and is to be explained in terms of, rule-governed processes, as opposed to the substance metaphysics view.

proof: A proof is a logical argument in which the truth of the statement to be proven is shown to follow necessarily from statements already accepted as true, hence, a deductive logical argument.

proteins: Complex, typically very large molecules made up of combinations of the 20 amino acids living cells manufacture, each protein possessing a distinctive spatial arrangement, or folding, of its components. Proteins, which determine cell metabolism, are manufactured on molecules called *ribosomes* in response to instructions from DNA via messenger RNA.

quantum chromodynamics (QCD): The quark theory of matter that developed in the 1960s in which hadrons (protons, neutrons, and all other particles that respond to the so-called strong nuclear force—thus, not electrons and neutrinos) are built up out of some combination of six quarks held together by gluons. Quarks and the electron-neutrino family of particles, called *leptons*, are now the *really* elementary forms of matter.

quantum electrodynamics (QED): The theory that developed in the 1940s out of Paul Dirac's 1929 quantum theory of the electromagnetic field. It describes the interaction of electrons, protons, and photons and is, thus, the quantum analogue of Maxwell's electromagnetic field theory.

rationalism: The view that deductive reasoning is both the only route to truth and capable of discovering all truths.

reverse engineering: Decomposing an artifact or a process in order to identify its components and mode of operation.

rhetoric: In ancient Greece, techniques of persuasive arguing that use the power of speech to win arguments, as opposed to the use of logical reasoning to prove the point.

science: Literally, knowledge, but for mainstream Western philosophy since Plato, it means knowledge that is universal, necessary, and certain because what we know is deducible from universal problems, not from individual facts.

semiconductor: A substance that is capable of being either a conductor of electricity or an insulator, depending on certain subtle and controllable changes in its makeup. Virtually all electronics technology since the 1960s has been based on semiconductor materials, especially silicon.

skeptics: Philosophers who deny the possibility of universal, necessary, and certain truths about nature; hence, they deny the very possibility of knowledge à la Plato-Aristotle and that anyone has achieved it.

spectroscope: An optical device that separates the many individual frequencies that make up the light incident upon it. Because the atoms of each element, when excited, radiate a distinctive set of frequencies, the elements in starlight and in laboratory specimens of matter can be identified.

spinning jenny: The name given to the single-operator–multi-spindle machine invented by James Hargreaves in the 1760s that revolutionized the production of cotton thread.

spontaneous symmetry breaking: An idea adopted by physicists in the 1960s to explain how a uniform state of affairs can evolve into a non-uniform one.

standard model: The name for the quantum theory that unified the electroweak theory and quantum chromodynamics, hence, the electromagnetic, weak, and strong forces. Since the 1970s, this theory has matured into our most powerful theory of matter and energy.

statistical mechanics: An extension of the ideas underlying the kinetic theory of gases by Boltzmann, Maxwell, and J. Willard Gibbs and developed further by others in the 20th century, including Einstein, for deriving observable properties of systems of material bodies from statistical models of the behavior of their parts.

statistical thermodynamics: The development of the kinetic theory of gases led Boltzmann and Maxwell to apply statistical models to the laws of thermodynamics and, thus, to energy flows.

stereochemistry: The study of molecular properties that derive from the spatial arrangement of the atoms in a molecule, as distinct from the properties that derive from the atomic composition of the molecule.

string theory: The name given to one approach to the final step in unifying the four fundamental forces in nature by uniting the standard model of quantum theory with the force of gravity, now described by the general theory of relativity, which is not a quantum theory. String theory proposes that all forms of matter and energy at all levels are variations on fundamental entities called *strings* that vibrate in 11-dimensional space-time.

substance metaphysics: The view, tracing back to the teachings of Parmenides, obscure even in antiquity, that the ultimate constituents of reality are timeless, changeless "things" with fixed properties, out of which all changing things are constructed and by means of which all change is to be explained.

syllabary: A writing system in which each symbol stands for a syllable in that language and the set of symbols/syllables allows the construction of all words in that language.

symbolic logic: A symbolic notation for recording and analyzing logical arguments and their properties, developed in the 19th century and leading to a revolution in logical theory and to the creation of a new discipline: mathematical logic.

symmetry: Initially descriptive of properties of mathematical forms, in the 19th and, especially, the 20th centuries, it was made into a fundamental principle of the physical world in physics, chemistry, and biology. In physics, symmetry plays a central role in so-called gauge theories, which attempt to unify the four fundamental forces in nature by supposing that they are the "debris" of symmetries that defined uniform forces in the very early history of the Universe but fragmented as the Universe cooled. See **spontaneous symmetry breaking**.

system: An ordered whole of mutually adapted parts keyed to the functionality of the whole.

taxonomy: A taxonomy is a systematic classification scheme that may be explicitly artificial, such as the familiar public library Dewey decimal system for classifying books (*not* created by the philosopher John Dewey!), or it may be natural. See **classification problem**.

techno-science: A name for technologies whose design and effective operation are dependent on scientific knowledge. Historically, technological innovations were quite independent of scientific theories, but this situation began to change in the 19th century. The commercial exploitation of these technologies led to the systematic coupling of science and engineering in industrial corporations, research labs, and academe.

temperament: See **tuning system**.

transistor: A device, invented at Bell Labs in 1948 by William Shockley, John Bardeen, and James Brattain, that exploits the properties of simple solid semiconductors to perform electronic functions then performed by much larger, less reliable, more expensive, and more power-consuming vacuum tubes.

tuning system: In music, a set of mathematical relationships among the notes of the octave that, when applied to the construction of musical instruments, maintains harmonies among the notes and minimizes dissonances. Tuning systems were inspired by Pythagoras's insight that mathematical relationships distinguish music from noise, but no one has discovered a single temperament or tuning system that is dissonance-free. Western music over the past 200 years employs equal temperament, a system in which the octave is divided into 12 equally spaced tones.

verisimilitude: Renaissance Humanist idea of the truthfulness of history writing that employs imaginative reconstructions and, by extension, the truthfulness of paintings that obviously are not what they seem to be.

water frame: A name given to the large water-powered version of Hargreaves's spinning jenny built by Richard Arkwright; together with related water-powered machinery for carding and weaving, the water frame initiated the mass-production era, setting the stage for the steam-power-based Industrial Revolution.

Biographical Notes

abu-Kamil (c. 850–930). Early Islamic algebraist, born in Egypt, and author of an influential text translated during the Renaissance containing 69 algebraic problems and their solutions.

al-Khwarizmi (c. 780–c. 850). The earliest known Islamic algebraist. His book of problems and solutions, along with that by abu-Kamil, influenced the shift of European mathematics from geometry to algebra.

Anaxagoras (c. 500–428 B.C.E.). Greek philosopher who proposed that all material objects were composed of a vast number of atoms of many different properties.

Archimedes (c. 287–212 B.C.E.). Greek mathematician and physicist whose combination of deduction and experiment influenced Galileo and Newton. He formulated a mathematics-based theory of the so-called simple machines and founded the science of hydrostatics.

Aristarchus of Samos (c. 320–c. 250 B.C.E.). Greek mathematician and astronomer who used trigonometry to estimate the distances to the Sun and Moon and proposed a Sun-centered Solar System that Copernicus read about in a text by Archimedes.

Aristotle (384–322 B.C.E.). Greek philosopher born in Macedonia, where his father was physician to the king. He studied with Plato for many years but then founded his own rival school, also in Athens. His comprehensive writings, especially on logic and nature, and his metaphysics, were extremely influential for more than 2000 years.

Avery, Oswald (1877–1945). American bacteriologist (born in Canada) who, together with Colin MacLeod and Maclyn McCarthy, argued in the early 1940s, based on their experiments with pneumonia bacteria, that DNA was responsible for inheritance.

Avicenna (980–1037). Islamic philosopher (Aristotelian) and physician, whose masterwork, *The Canon of Medicine*, became a standard text, alongside Galen's, for more than 500 years in European medical schools.

Bacon, Francis (1561–1626). English educational reformer; also reformer and "father" of the experimental method in modern science described in his book *The New Organon* (1620); and political opportunist. Became Lord High Chancellor under King James but was convicted of bribery.

Bernard of Clairvaux (1090–1153). Extremely influential 12th-century French theologian and Church leader, head of the Cistercian order of monasteries that extended over much of Europe. He opposed the rising secular intellectualism that became institutionalized in the emerging universities and especially persecuted the philosopher Peter Abelard.

Bernard, Claude (1813–1878). French biologist, founder of experimental medicine, and an extremely prolific author of research publications, many of whose results remain valid today. He was a positivist, favoring facts over concepts, and championed a homeostatic view of metabolism.

Bernoulli, Jacob (1654–1705). Swiss mathematician, member of a family of outstanding mathematicians in the 17th and 18th centuries, whose posthumously published *The Art of Conjecturing* pioneered probability theory and its application to political and commercial decision-making.

Bohr, Niels (1885–1962). Danish physicist, deeply philosophical as well, whose 1913 proposal to quantize the orbital motion of electrons became the foundation of quantum mechanics. In the late 1920s, with Werner Heisenberg, he formulated the Copenhagen Interpretation of quantum mechanics.

Boltzmann, Ludwig (1844–1906). Austrian physicist who founded statistical mechanics and statistical thermodynamics, posited the kinetic theory of gases (with James Clerk Maxwell), and insisted on the physical reality of atoms.

Boole, George (1815–1864). British mathematician who founded modern mathematical logic by introducing a symbolic notation for logical reasoning. His 1854 book, *An Investigation into the Laws of Thought*, was of immense influence in the history of information theory, computers, and artificial intelligence research, as well as in mathematical logic.

Boyle, Robert (1627–1691). Irish natural philosopher, heir to the title earl of Cork, and member of the Oxford group of natural philosophers that founded the Royal Society of London. Boyle conducted experiments with Robert Hooke, using an air pump of their design, on the physical properties of air; these were considered exemplary of the experimental method of the study of nature. He was an atomist and an early "scientific" chemist.

Brahe, Tycho (1546–1601). A Danish astronomer, the greatest observational astronomer of the pre-telescope era, who used instruments of his own design in an observatory funded by the Danish king. He rejected Copernicus's theory for his own version of an Earth-centered Universe. His data were used by Johannes Kepler to support Kepler's claim that the planets move in elliptical orbits.

Brunelleschi, Filippo (1377–1446). Italian painter, sculptor, and architect; famous for his rediscovery of perspective drawing and his innovative design and construction plans for the cathedral in Florence with its vast dome and cupola.

Bush, Vannevar (1890–1974). American engineer, science administrator, head of the World War II Office of Scientific Research and Development, author of the report that launched large-scale postwar federal support for research, and computer pioneer.

Cardano, Jerome (Girolamo) (1501–1576). Italian mathematician, physician, and founder of probability theory applied to gambling games. He promoted the study of algebra and published Tartaglia's solution to the cubic equation.

Clausius, Rudolf (1822–1888). German physicist and a founder of thermodynamics, Clausius introduced the concept of entropy, implying the irreversibility of time and the "heat death" of the Universe. With Maxwell and Boltzmann, he also created the kinetic theory of gases.

Comte, Auguste (1798–1857). French social and political philosopher and philosopher of science; founder of positivism—basing knowledge on facts, not ideas—and of sociology.

Copernicus, Nicolaus (1473–1543). Polish astronomer and physician whose theory of a moving Earth eventually redirected astronomy. Copernicus spent years studying in Italy after graduating from Jagiellonian University in Krakow and became proficient in Greek, translating into Latin the work of an ancient Greek poet recovered by the Humanists. It is interesting that Copernicus used virtually the same data that Ptolemy used yet reached dramatically different conclusions.

Crick, Francis (1916–2004). English physicist. After working on radar and magnetic mines during World War II, Crick collaborated with James Watson on the spatial arrangement of the atoms in DNA molecules; the two shared the Nobel Prize for that 1953 discovery.

Ctesibius (3rd century B.C.E.?). A Greek "mechanic," son of a barber, who invented a wide range of useful machines, including a complex water clock and a water organ, that were developed further by others over the next 300 years.

Curie, Marie (1867–1934). Born in Warsaw, Curie moved to Paris with a newly married older sister in 1891 and married Pierre Curie in 1895. They shared a Nobel Prize in physics in 1903 with Henri Becquerel for the discovery of radioactivity, named by Marie in 1898. After Pierre's death in 1906, Marie became the first woman professor at the Sorbonne, and in 1911, she was awarded the Nobel Prize in chemistry for her isolation of radium.

Dalton, John (1766–1844). Not the first atomist of modern times—Boyle and Newton were among his many predecessors—Dalton's *New System of Chemical Philosophy* (1807) became the foundation of 19th-century atomic theories of matter, first in chemistry, later in physics.

Darwin, Charles (1809–1882). Born into a wealthy English family, Darwin married a cousin, Emma Wedgwood, and devoted his life to biological science. In addition to his theory of evolution, Darwin published extensively on many subjects and would be considered a major 19th-century scientist independent of evolution.

Darwin, Erasmus (1731–1802). Charles Darwin's paternal grandfather and author of several once-popular (although later mocked) epic poems on nature that incorporated evolutionary ideas of his own.

Davy, Humphrey (1778–1829). An English physicist/chemist, Davy was extraordinarily productive in both "pure" and applied research, pioneering electrochemistry and discovering the elements sodium and potassium, as well as inventing a safety lamp for coal miners, an electric arc lamp, and a process for desalinating seawater.

Dee, John (1527–1608/09). English mathematician and mystical nature philosopher. Dee was actively involved in training ship pilots in the new mathematical techniques of navigation and in mathematical cartography, as well as promoting mathematics literacy for the public. He designed "magical" stage machinery for plays and made the first translation into English of Euclid's *Elements*.

Descartes, René (1596–1650). Descartes was a founder of modern science, modern philosophy, and modern mathematics. He promoted a deductive method for acquiring knowledge of nature and developed a rigorously mechanical philosophy of nature in which only contact forces were allowed: no action at a distance. He made epistemology (the theory of knowledge) central to philosophy, and he invented analytic geometry, making algebra central to mathematics.

De Vries, Hugo (1848–1935). Perhaps the leading figure in founding modern genetics, De Vries, a Dutch botanist, rediscovered Mendel's ignored earlier work after developing his own similar theory and gave the credit to Mendel. He developed an influential theory of mutations as the "engine" of evolution.

Dirac, Paul (1902–1984). Trained initially as an engineer at Bristol University in England, Dirac became one of the greatest theoretical physicists of the 20th century. His 1929 relativistic theory of the electron became the cornerstone of quantum electrodynamics, the most important theory in physics in the mid-20th century.

Dolland, John (1706–1761). An English weaver by training, Dolland became a self-educated scientist, developing and patenting the first compound microscope lenses corrected for chromatic aberration.

Dumas, Jean Baptiste (1800–1884). A French chemist who developed a technique for calculating relative atomic weights, Dumas also pioneered structuralism in chemistry through his theory of substitution of atoms in geometric "types" of molecules.

Einstein, Albert (1879–1955). Given all that has been written about him, perhaps the most amazing fact about Einstein is that, in 1904, no one, with the exception of his closest friend and sounding board, Marcel Grossmann, would have predicted his subsequent accomplishments. In spite of his epochal 1905 papers, his reputation flowered only from 1911. He was appointed director of the Kaiser Wilhelm Institute for Physics in 1914 and, in 1915, published the general theory of relativity. He resigned in 1933 and settled at the Institute for Advanced Studies in Princeton, which was created in part to provide a "home" for him.

Empedocles (c. 490–430 B.C.E.). An early Greek natural philosopher who formulated a four-element theory of matter—earth, air, fire, water—that, together with attractive and repulsive forces, lasted into the 18th century.

Epicurus (341–270 B.C.E.). A Greek moral philosopher primarily, who adopted Democritus's atomic theory of matter and adapted it to his moral and social views. Against Anaxagoras, he held that atoms differed only in size, shape, and weight and that all properties of material objects derived from diverse configurations of their constituent atoms.

Erasmus of Rotterdam (1466–1536). One of the great Humanist scholars and the first author, it is said, to live wholly off his fees from publishers based on the sale of his books, especially his bestselling *In Praise of Folly*.

Euclid (c. 300 B.C.E.). A Greek mathematician, whose synthesis of 200 years of Greek mathematics into an axiomatic system in his book, *The Elements*, was of incalculable influence in Western philosophy, mathematics, and science, right down to the present day. Almost nothing is known of his personal life.

Euler, Leonhard (1707–1783). One of the greatest mathematicians of all time and, perhaps, the most productive. Born in Basel, he lived most of his adult life in Germany or Russia, writing on pure and applied mathematical problems even after he became blind. He encompassed all of mathematics but contributed especially to "analysis," another name for algebra, and made important contributions to astronomy, optics, mechanics, and engineering mechanics: the rigorous solution of engineering problems.

Faraday, Michael (1791–1867). A gifted and highly prolific experimental physicist and chemist, Faraday was effectively wholly self-educated, though he was never proficient in mathematics. He became Humphrey Davy's assistant at the Royal Institution in London through an accident, and he was later Davy's successor. He discovered the dynamo principle in 1830, and invented the concepts of electric and magnetic fields and lines of force, predicting that light was an electromagnetic phenomenon. He rejected the atomic theory.

Fischer, Emil (1852–1919). A German organic chemist famous, first, for his synthesis of sugars and, later, for synthesizing amino acids, then combining them to form proteins. His lock-and-key metaphor for how enzymes act on cell molecules was and remains a powerful heuristic in molecular biology.

Fourier, Joseph (1768–1830). French mathematical physicist whose *Analytical Theory of Heat* was extremely influential, both in terms of its equations and in separating descriptive physics from metaphysics. His use of simple trigonometric functions to model any periodic behavior, however complex, remains one of the most powerful tools in science and engineering.

Francesca, Pierro della (1420–1492). One of the great Renaissance painters, he was also a mathematician and wrote perhaps the first account of perspective drawing as a mathematical technique, *De prospectiva pingendi*. This then became a staple of 15th-century artist's manuals, especially after Leone Alberti's influential *Della pittura* (1436).

Galen (c. 129–199). Greek physician and medical theorist, prolific writer and experimenter, and physician to various Roman emperors. Galen was to medieval and Renaissance medicine what Aristotle was to philosophy: the authority. His theory of health as a balance among four humors was influential into the 19th century, though his anatomy and physiology were overthrown in the 16th and 17th centuries.

Galilei, Galileo (1564–1642). Italian mathematical physicist and founding "father" of modern science, combining deductive reasoning and extensive experimentation à la Archimedes. Born in Pisa, he became a professor of mathematics first there, then in Padua, after his telescope-based observations of the Moon's irregular surface and Jupiter's moons made him famous. His condemnation for teaching Copernicus's theory as true came in 1633.

Galilei, Vincenzo (1520–1591). Galileo's father; Vincenzo was a musician and a music theorist at a time of intense controversy over tuning systems and their mathematical models. He broke with his teacher, Zarlino, who defended an expanded Pythagorean system, in favor of equal-temperament tuning. Vincenzo's books reveal clever experimentation to support his claims.

Gamow, George (1904–1968). Born in Russia, Gamow moved west after earning his Ph.D., studying the new quantum physics first in Germany, then with Bohr in Copenhagen, before settling in the United States. He predicted the quantum tunneling effect in 1929; proposed the Big Bang theory of cosmology in the late 1940s, predicting the microwave background radiation detected in 1963; and proposed that the sequence of bases in the Watson-Crick DNA model was a code for producing proteins out of amino acids.

Gutenberg, Johann (c. 1397–1468). Widely but by no means unanimously considered the inventor of movable-metal-type printing. Almost nothing about Gutenberg's life and work is free of uncertainty except that he was born in Mainz on the Rhine River, apprenticed as a goldsmith but became a printer, and printed a number of deluxe copies of the Bible using metal type in the early or mid-1450s.

Guth, Alan (1947–). American physicist who, in 1980, proposed the inflation model of the Universe, preceding Gamow's Big Bang. Subsequent refinement by others, as well as by Guth, and detailed observation of the microwave background radiation's minute non-uniformities led to a consensus in favor of the inflation model.

Hegel, G. F. W. (1770–1831). A German philosopher, Hegel was the single most influential philosopher of the 19th century, the creator of a system that integrated deduction, history, and time. He held that reality is the deterministic unfolding in time of reason, which manifests itself as nature and as the human mind.

Heisenberg, Werner (1901–1976). A German physicist, Heisenberg invented, in 1925, what was later called "quantum mechanics." Over the next five years, he formulated the famous uncertainty principle and, in collaboration with Bohr, an interpretation of quantum theory that was probabilistic and strictly empirical. Bohr broke with Heisenberg over the latter's role as head of Germany's wartime atomic bomb research effort.

Helmholtz, Hermann (1821–1894). A physicist and pioneering neurophysiologist, Helmholtz was Germany's leading scientist in the second half of the 19th century. He formulated the scientific principle of the conservation of energy, studied the transmission of signals in nerves, and developed a theory of hearing that became the basis for designing stereo audio equipment.

Heraclitus (c. 540– c. 480 B.C.E.). Other than that he lived in the Greek city of Ephesus in what is now Turkey and probably wrote before Parmenides, not after, nothing is known about Heraclitus. That he wrote one or more works on philosophy is known, and in these, he clearly insisted on the reality of change, suggesting that the object of knowledge is the *logos*, or orderliness, of processes, not timeless objects and their properties.

Hero (or Heron) of Alexandria (flourished c. 60). A Greek "engineer" before the term existed, Heron created a school for engineering in Alexandria and left behind a number of books describing mechanical and optical machines based on physical principles, including the action of compressed air, water, and steam.

Hertz, Heinrich (1857–1894). A German physicist who, in the 1880s and independently of Oliver Lodge in England, confirmed the prediction of Maxwell's theory of electromagnetic waves traveling freely through space. This finding became the basis for the broadcast radio technology developed 10 years later by Guglielmo Marconi.

Hippocrates (c. 460–377 B.C.E.). A Greek physician, medical theorist, founder of a medical school, and teacher. His school was on the island of Kos, where he was born, and pioneered a wholly naturalistic approach to illness and treatment.

Hoffman, August (1818–1892). An eminent German organic chemist, Hoffman was called to London by Prince Albert to teach at the new Royal College of Chemistry, where his student William Perkin synthesized the first artificial dye from coal tar. Hoffman later returned to Germany, founded the German Chemical Society, and played an active role in German chemists' dominance of the commercial dye industry and the many important industrial applications of coal-tar chemistry.

Holland, John H. (1929–). American computer scientist and creator of "genetic" algorithms: computer programs based on Darwinian evolution and genetic theory that display adaptation. Holland is a theorist of complex systems and self-organization and was actively involved with the Santa Fe Institute and World Economic Forum, in addition to teaching computer science at the University of Michigan.

Hooke, Robert (1635–1703). An English natural philosopher, Hooke collaborated with Robert Boyle on experiments to determine the properties of air, studied the properties of metallic springs, and invented a spiral spring-controlled balance wheel for a watch (replacing the pendulum). He was a pioneering microscopist, invented numerous scientific instruments, attempted a theory of gravity, and played a leading role in the rebuilding of London after the great fire of 1665.

Hoyle, Fred (1915–2001). An English physicist, Hoyle, together with Herman Bondi and Thomas Gold, proposed the Steady State theory of the Universe as a counter to Gamow's Big Bang theory, a name mockingly assigned by Hoyle. Hoyle also wrote science fiction and argued that life came to Earth from space.

Hubble, Edwin (1889–1953). Hubble was a midwestern American astronomer who became director of the Mt. Wilson observatory in 1919 and, with its 100-inch reflecting telescope, soon discovered that the sky was filled with galaxies, contrary to the consensus view that the Milky Way was the Universe. In 1929, Hubble announced the expansion of the Universe and devoted the rest of his life to observations aimed at determining its size and age.

Huygens, Christiaan (1629–1675). A Dutch mathematical physicist, mathematician, and astronomer, Huygens was a central figure in the creation of modern science, first to demonstrate that curved motion required a force and to recognize Saturn's rings as such. He developed a wave theory of light; made important contributions to algebra, probability theory, optics, and mechanics; developed accurate pendulum clocks and their theory; and independently of Hooke, invented a spring-balance wheel-driven watch.

Joule, James Prescott (1818–1889). An English physicist whose experiments on the quantitative relationship of mechanical motion and heat led, in the hands of others, to the idea of conservation of energy and the creation of thermodynamics.

Kekule, Friedrich August (1829–1886). A German organic chemist, Kekule is best known for his contributions to structural chemistry, especially the hexagonal ring structure of benzene and his prediction of the tetrahedral form of the carbon atom's valence bonds, which became the basis of polymer chemistry in the 20th century.

Kelvin, Lord/William Thomson (1824–1907). An Irish-born mathematical physicist, Thomson was knighted for designing and overseeing the laying of the first successful transatlantic telegraph cable in 1866. He played key roles in the development of thermodynamics and electromagnetic field theory but was wedded to the reality of the aether and believed that the Earth was probably only 100 million years old and, thus, too young for Darwin's theory of evolution to be correct.

Kepler, Johannes (1571–1630). A German astronomer who first formulated the modern conception of the Solar System, which is very different from that of Copernicus. The data Kepler used came from Tycho Brahe, whose assistant he became when Brahe relocated to Prague. When Brahe died, Kepler took the data and applied them, first, to a Pythagorean theory of his own that failed to match the data; he then let the data guide his theorizing, arriving at elliptical orbits, not circular ones.

Khayyam, Umar (1048–1131). Islamic mathematician, astronomer, and poet who had effectively achieved a general solution to the cubic equation centuries before Tartaglia, and whose text *Algebra* anticipates Descartes' invention of analytic geometry.

Koch, Robert (1843–1910). A German biologist who, with Louis Pasteur, founded bacteriology and formulated the germ theory of disease. His Nobel Prize was for discovering the bacterium that causes tuberculosis, then rampant, but he developed methodologies for isolating microorganisms that resulted in his discovery of anthrax and cholera bacteria, as well.

Lamarck, Jean-Baptiste (1744–1829). Lamarck was almost 50, with very modest credentials as a botanist, when in 1793 the committee running the French revolutionary government made him a national professor of invertebrate zoology. His theory of the emergence of all life forms from a common ancestor by natural forces was an important predecessor of Charles Darwin's theory, and one of which Darwin was acutely aware.

Laplace, Pierre-Simon (1749–1827). A French mathematical physicist and theoretical astronomer of great influence who managed to prosper under Louis XVI, the revolutionary government, Napoleon, and the restored Bourbon monarchy! He proved the long-term stability of the Solar System under Newtonian gravitation, developed a mathematical theory of the origin of the Solar System out of a cloud of gas, published an important essay on probabilities, and championed a rigorous materialistic determinism.

Laurent, Auguste (1807–1853). At one time a graduate student of Dumas (above) contemporary with Pasteur, Laurent seems first to have developed a theory that the spatial arrangement of atoms in a molecule determines properties of the molecule. His "nucleus" theory was subsequently overwhelmed by Dumas' extension and adaptation of it, a source of some bitterness to Laurent.

Lavoisier, Antoine de (1743–1794). Unlike Laplace, Lavoisier did not survive the French Revolution. His research led him to believe that combustion involved combination with one component of air, which he named *oxygen*. This led him to propose a "revolution" in chemistry, one that laid the foundation for the modern theory of elements. His widow married Benjamin Thompson (below).

Leavitt, Henrietta Swan (1868–1921). An American astronomer, Leavitt graduated from what became Radcliffe College and became a human "computer" at Harvard College Observatory, eventually rising to head of a department there. Her specialty was variable stars, and she identified 2400 new ones, especially the Cepheid variables that she recognized as providing a cosmic "ruler" for measuring absolute cosmic distances.

Leibniz, Gottfried (1646–1716). A German philosopher, mathematician, and physicist, Leibniz, like Descartes, was influential in all three of those areas. He formulated a rationalist, deterministic but anti-materialistic philosophy; invented the calculus independently of Newton (publishing first), using a notation that has become universal; anticipated late-19th-century topology and symbolic logic; and first called attention to the quantity in mechanics that we call kinetic energy.

Liebig, Justus von (1803–1873). A German chemist of enormous influence, partly through his own mechanistic theories of chemical reactions, but largely through the many subsequently prominent students trained in his laboratory. Liebig studied the chemistry of fermentation long before Pasteur and never accepted that the cause was a living organism (yeast). He also dismissed the significance of atomic structure within molecules and the early germ theory of disease.

Linnaeus, Carl (1707–1778). Swedish botanist whose binomial system for classifying plants based on their sexual organs became universally adopted. An aggressive proponent of his system as a natural one, he was forced to acknowledge late in life that it seemed to be conventional, which implied that species and genera were conventional, not immutable features of reality.

Lucretius (c. 99/94–c. 55/49 B.C.E.). Roman poet and natural philosopher whose epic poem in hexameters, *On the Nature of Things*, disseminated Epicurus's atomic theory of matter and morality, somewhat modified by Lucretius.

Mach, Ernst (1838–1916). An Austrian experimental physicist of note but remembered mostly for his theory of scientific knowledge as based on perceptual experience and incapable of penetrating to a reality behind experience, which is why he opposed the reality of atoms.

Maxwell, James Clerk (1831–1879). One of the greatest of all mathematical physicists, Maxwell was born in Edinburgh. He became, in

1871, the first professor of experimental physics at Cambridge University and established the Cavendish Laboratory there. Under the leadership of Lord Rayleigh, J. J. Thomson, and Ernest Rutherford, the laboratory became a leading center for important new developments in physics into the 1950s.

Mendel, Gregor Johann (1822–1884). An Austrian monk and botanist, Mendel lived in the Augustinian monastery in Brünn (Brno) effectively for the last 40 years of his life, becoming abbot in 1868, which ended his experimental work on plants. Mendel twice failed to pass the exams for an advanced teaching license. In between failures, he spent three years at the University of Vienna studying science and mathematics and, on returning to the monastery in 1854, began the years-long breeding experiments that resulted in his posthumous fame.

Mercator, Gerard (1512–1594). Mercator was born in Flanders and, after a religious crisis that led to his becoming a Protestant, studied mathematics in order to apply it to geography and cartography. He migrated to the Lutheran town of Duisberg in Germany in 1552, living there for the rest of his life and producing the first modern maps of Europe over the next 10–15 years. In 1569, he produced a map of the Earth based on his projection of its surface onto the inner surface of a cylinder. He coined the term *atlas* for a collection of maps.

Morgan, Thomas Hunt (1866–1945). An American geneticist, Morgan began his career as an embryologist, studying fertilization at the cell level. In 1907, after becoming a professor at Columbia University, he shifted his research to the mechanism of heredity. Initially critical of the gene concept, he became a major proponent of it after 1910 and trained a number of influential students, among them, Hermann Muller (below).

Morse, Samuel F. B. (1791–1872). Morse was a financially unsuccessful American artist who, returning from Europe in 1832 after a three-year stay, learned about electromagnetism from a fellow passenger. Morse became obsessed with the idea of an electric telegraph and, eventually, with advice from the physicist Joseph Henry, among others, succeeded in getting Congress to fund a pilot line from Baltimore to Washington in 1843. This inaugurated the commercialization of the telegraph using Morse's code.

Muller, Hermann Joseph (1890–1967). Born in New York City, Muller attended Columbia University and received his Ph.D. under Thomas Hunt Morgan's direction in 1916, by which time he was Morgan's active collaborator in research and publication. His Nobel Prize–winning discovery of genetic mutations induced by X-rays was made while he was at the University of Texas, but he spent the mid-1930s at the Institute of Genetics in the Soviet Union, leaving because of his opposition to Lysenko's anti-Mendelian theories, which were approved by Stalin. He returned to the United States in 1940.

Müller, Johannes (1801–1858). Müller was a German physiologist who was committed to the vitalist view of life and to Romantic nature philosophy, yet his research laboratory, first at the University in Bonn, then in Berlin, was the training ground for an extraordinary group of influential life scientists, virtually all of whom were mechanists!

Newton, Isaac (1642–1727). Great both as an experimental and as a theoretical physicist, Newton's "miraculous year" was 1665, when an outbreak of plague caused Cambridge University to close and he went home. In notebooks he kept then, there are the clear antecedents of most of his great ideas in mechanics, optics, mathematics, and astronomy. Newton devoted much of his time (and the bulk of his surviving writing) to biblical chronology and interpretation and alchemical researches, yet he was the single most important architect of modern science. He was autocratic as warden of the Royal Mint and as president of the Royal Society, and suffered a mental breakdown in the early 1690s, perhaps poisoned by his alchemical experiments.

Pacioli, Luca (1445–1517). Renaissance mathematician, befriended by Piero della Francesca, and tutor to, and friend of, Leonardo da Vinci. Pacioli published a summary of 15th-century mathematics in 1494 that included an extensive description of double-entry bookkeeping and commercial arithmetic generally. His 1509 book, *On the Divine Proportion* (the famous Greek "golden ratio"), was written in Italian and illustrated by Leonardo. It describes the application of mathematical proportions to artistic depictions, for example, of the human body.

Parmenides (born c. 515 B.C.E.). One of the earliest and most influential Greek philosophers, in spite of the fact that his only work is lost, a poem of some 3000 lines, apparently, of which 150 are known because they are cited by other Greek philosophers. Parmenides's rigorously logical characterization of the concepts of being and becoming provoked atomistic theories of nature, in contrast to Heraclitus's process approach and influenced the view, dominant since Plato and Aristotle, that reality was timeless and unchanging and that knowledge and truth were universal, necessary, and certain.

Pascal, Blaise (1623–1662). A French mathematician and physicist, in 1654, Pascal had a vision that led him to cease almost all secular intellectual activity, although he designed a public transportation system for Paris that was built the year he died. He made important contributions to projective geometry and probability theory before turning to philosophical and theological themes. His *Pensées* has been in print since publication.

Pasteur, Louis (1822–1895). A French chemist, Pasteur became the very embodiment of the natural scientist, for the French at least. With Robert Koch, he formulated the germ theory of disease, but he also contributed to the creation of stereochemistry and established the value of chemical science to industry through his work on fermentation, pasteurization, vaccination, and silkworms.

Petrarch (1304–1374). An Italian poet, born in Arezzo, Petrarch was named Poet Laureate of Rome in 1341, largely because of an epic poem in Latin on the great Roman general Scipio Africanus, who defeated Hannibal. He was a great admirer of Dante, whose *Comedy* (called "divine" by Petrarch's admirer Boccaccio) was in Italian, not Latin. Petrarch instigated the Humanist movement and the collection of ancient manuscripts in order to recover models of the best writing, feeling, and thinking.

Planck, Max (1858–1947). The German physicist who initiated the quantum physics "revolution" with his 1900 solution to the black-body radiation problem. Planck remained in Germany in spite of his outspoken opposition to Nazi policies, and his only surviving son was gruesomely executed as an accomplice to an assassination plot on Hitler. After World War II, the Kaiser Wilhelm Institute was renamed the Max Planck Institute(s).

Plato (428–347 B.C.E.). The quintessential Greek philosopher from the perspective of the subsequent history of Western philosophy, Plato was one of Socrates's students and the teacher of Aristotle. "Plato" was his nickname—his given name was Aristocles—and he was initially a poet and, as a young man, a competitive wrestler. He was an elitist by birth and inclination, and the relationship between the "real" Socrates and the character in Plato's dialogues is, at best, loose.

Prigogine, Ilya (1917–2003). A Russian-born chemist who was raised, educated, and rose to fame in Belgium. He moved to the United States in 1961, first to the University of Chicago, then to the University of Texas, while retaining his Belgian academic affiliation. He demonstrated that many far-from-equilibrium physical and biological systems were self-organizing and stable and displayed adaptation.

Pythagoras (c. 572–c. 479 B.C.E.). A Greek philosopher and mathematician with a strong metaphysical/mystical bent. Pythagoras promoted a total lifestyle conception of wisdom and created schools and self-contained communities in which people could live in accordance with his teachings. His most enduring accomplishments are the idea of deductive proof, which essentially created mathematics as we know it, and the idea that mathematical forms are the basis of all natural order.

Quetelet, Adolphe (1796–1874). A Belgian astronomer-in-training whose lasting achievement was social statistics, especially the idea of statistical laws, which challenged the prevailing belief that laws were necessarily and exclusively deterministic.

Rumford, Count/Benjamin Thompson (1753–1814). A Royalist, Thompson fled the colonies to England in 1776, returning to the colonies as a British officer, then went to Europe after the war. He distinguished himself in Bavaria as minister of war and minister of police, becoming de facto prime minister, and was made count of the Holy Roman Empire in 1791. He instituted workhouses for the poor and new uniforms, marching formations, diet, and weapons for the army. In addition to his experiment in 1799, which proved that heat was motion, he founded the Royal Institution in London to teach science to the public, enjoyed a short-lived marriage to Lavoisier's widow, and endowed a science professorship at Harvard.

Rutherford, Ernest (1871–1937). Born in New Zealand, Rutherford went to Cambridge in 1895 to study with J. J. Thomson, then to McGill in Montreal as professor of physics, before returning for good to England in 1907. He was professor of physics in Manchester until 1919 when he moved to the Cavendish Laboratory at Cambridge as J.J. Thomson's successor. His 1908 Nobel Prize was in chemistry for his work on radioactivity, but he and his students made many fundamental contributions to atomic and nuclear physics.

Schleiden, Mathias (1804–1881). A German botanist who, with Theodor Schwann, proposed the cell theory of life. Schleiden was originally a lawyer who turned to the study of botany after a failed suicide attempt. In 1838, he published an essay in a journal edited by Johannes Müller, proposing that cells are the basis of all plant life and are formed in a process that begins inside the nucleus of a progenitor cell. In an 1842 text, he argued that a single, mathematically describable physical force underlies all natural phenomena, including life.

Schrödinger, Erwin (1887–1961). Schrödinger was born in Vienna and received his Ph.D. there, in physics. He was an Austrian artillery officer during World War I and became a professor of physics in Zurich in 1921. It was at Zurich in 1925 that he developed his version of what was called "quantum mechanics," which unlike Heisenberg's version, was based on 19th-century deterministic wave physics and was interpretable as offering a conceptual "picture" of microphysical nature. Succeeding Max Planck as professor of theoretical physics in Berlin in 1927, he fled the Nazi takeover in 1933 for London; he returned to Vienna in 1936, only to flee again, this time, to Ireland until 1956 and yet another return to Vienna.

Schwann, Theodor (1810–1882). Schwann was born and educated in Germany and served from 1834–1838 as Johannes Müller's laboratory assistant in Berlin, but after a paper on yeast as a factor in fermentation was derided, he accepted a professorship in Belgium, where he spent his entire academic career. It was his 1839 book, *Microscopical Researches on the Similarity in the Structure and Growth of Animals and Plants*, that proposed the cell theory as the universal basis of both plants and animals, hence, of all life forms. Schwann thus extended Schleiden's cell theory of plant life, with which he was quite familiar; for this reason, the universal cell theory is attributed to them jointly.

Shannon, Claude (1916–2001). Shannon was born in Michigan and did his graduate work at MIT, receiving his master's and Ph.D. in mathematics in 1940 (with these in two different areas of applied mathematics). He joined Bell Labs in 1941. During the war, he worked on mathematical models for predictive antiaircraft firing. The technological applications of Shannon's theories in computer logic circuit design, telephone switching circuits, computer networks, and electronic and optical information transmission and storage devices have had an incalculable social impact.

Shapley, Harlow (1885–1972). An American astronomer, Shapley originally planned to become a journalist after attending a new university journalism program in his home state of Missouri. He became an astronomer because the program was delayed and he chose to take astronomy courses while he waited! At Princeton from 1911 to 1914, Shapley did important observational work on double stars and variable stars and was, thus, well positioned, after his move to Mt. Wilson Observatory in 1914, to use Leavitt's variable-star–based cosmic "ruler" to estimate the size of the Milky Way and distances to the Magellanic Clouds.

Swift, Gustavus (1839–1903). Born and raised on Cape Cod, Swift dropped out of school after eighth grade and, at the age of 16, had his own butchering business, which prospered and expanded with each relocation. He and his partner moved to Chicago in 1875, and in 1878, he went his own way, commissioning the first successful refrigerated rail car. It was delivered in 1880, and a year later, he had 200 cars carrying 3000 dressed beef carcasses a week to New York City.

Tartaglia, Niccolò (aka Niccolò Fontana) (1499–1557). Tartaglia was a mathematician and engineer; as a boy, he was slashed through the cheek by a French knight, hence, his nickname, "stammerer" ("Tartaglia"). He discovered the general solution to the cubic equation (though Khayyam may have anticipated him, independently) and the parabolic trajectory of cannonballs. He made the first Italian translations of Euclid and Archimedes.

Thales (c. 624/620–c. 570/546 B.C.E.). According to Aristotle, Thales was the first Greek natural philosopher, speculating that water or some fluid was the universal "stuff" out of which all material objects were composed.

Thompson, Benjamin. See **Count Rumford**.

Thompson, Joseph John (1856–1940). Thompson was born near Manchester in England and planned on studying engineering, but because the family could not afford the apprenticeship fees for engineering training, he switched to physics, winning a scholarship to Cambridge—where he spent the rest of his life—and a Nobel Prize (in 1906). In 1884, he succeeded Lord Rayleigh, who had succeeded Maxwell, as Cavendish Professor of Physics and was succeeded in turn by Ernest Rutherford when he resigned in 1919 to become Master of Trinity College. His Nobel Prize was for discovering the electron and for studies of the conduction of electricity in gases, establishing the electron as a particle. (Thompson's son George won the Nobel Prize for physics in 1937 for showing that electrons behaved like waves!)

Thomson, William. See **Lord Kelvin**.

Virchow, Rudolf (1821–1902). A German biologist and another student of Johannes Müller's, Virchow focused his research on the cell, especially cellular pathologies, which he believed to be the basis of all disease. In the mid-1850s, he insisted on the principle that cells arise only from other cells by a process of division, utterly rejecting the possibility of spontaneous generation or other cell formation theories current at the time, among them, Schleiden's. His text *Cellular Pathology* was influential, and its ideas were the basis of his rejection of the Pasteur-Koch germ theory of disease. Like most biologists prior to the mid-1870s, Virchow defended the inheritance of acquired characteristics.

Vitruvius (fl. 1st century B.C.E.). Almost nothing is known about Vitruvius's personal life (even his full name is conjectural!), but the influence of his *On Architecture* since the Renaissance has been immense. The book was discovered in 1414 by the Humanist scholar Poggio Bracciolini and popularized by Leone Alberti in a major work on art and architecture, *On the Art of Building* (1452), that was dedicated to Pope Nicholas V, who initiated the construction of the Vatican complex.

Wallace, Alfred Russel (1823–1913). Wallace was born into a poor Welsh family and left school at 14 to become a surveyor under an older brother. Self-educated by local standards, he became a teacher at a workingman's school in 1844 when he met a young naturalist, Henry Bates, and discovered his true vocation. They traveled to Brazil together in 1848 to gather specimens for wealthy British collectors, and Wallace returned to England with crates' worth, all of which were lost when the ship sank approaching the English coast. He then spent 20 years, from 1842–1862, in the Malay Peninsula, collecting specimens and studying the distribution of plants and animals; this led to the essay proposing evolution by natural selection that he sent to Darwin in 1858 for his advice as to publication. Wallace became one of England's greatest naturalists, but he never accepted the extension of evolution to man. He was an aggressive supporter of radical social reform and of "scientific" spiritualism, believing in life after death.

Watson, James (1928–). Watson studied zoology at the University of Chicago and received his Ph.D. from Indiana University in 1950, where he was influenced by T.H. Morgan's former student Hermann Muller. Watson's Ph.D. thesis, under Salvador Luria, was on the effect of X-rays on bacterial cell division. In 1951, he met Maurice Wilkins, who had with him copies of X-ray diffraction patterns of DNA crystals, and Watson decided to spend time at the Cavendish Laboratory at Cambridge studying them. There he met Francis Crick and began their epochal collaboration.

Watt, James (1736–1819). A Scottish mechanic and instrument maker who opened a shop in Glasgow circa 1756 and began working for faculty members at Glasgow University. While repairing a Newcomen engine, Watt saw that the efficiency would be improved dramatically by separating the cylinder and the condenser. He built a crude working model in 1765, but a reliable engine for commercial application required solving problems involving valves, precision machining (for the day), and lubricants. Watt's partnership with Mathew Bolton from 1774 was heaven-sent, initiating the steam-power–based Industrial Revolution of the 19th century and freeing Watt to develop progressively more efficient and more useful designs that Bolton put into production.

Bibliography

Agricola, Georgius (Georg Bauer). *De Re Metallica*. Herbert Hoover, trans. New York: Dover, 1950. Bauer was a sort of chief medical officer for the Fugger mines in the 16th century, and this is a translation—by Herbert Hoover before he became president—of his survey of contemporary technologies related to mining, including medical technologies, of course.

Allen, Garland. *Life Science in the Twentieth Century*. Cambridge: Cambridge University, 1981. A solid history of biology in the first half of the 20th century.

Aristotle. *The Complete Works of Aristotle*. Jonathan Barnes, ed. Princeton: Bollinger, 1982. Aristotle's logical works include the *Prior* and *Posterior Analytics* and the *Rhetoric*. His scientific works include *On the Heavens*, *On Physics* (nature), *On the Soul*, *On Generation and Corruption*, and the *Generation of Animals*.

Augustine. *Confessions*. London: Penguin Books, 1961. Book 11 contains his penetrating analysis of time, and his style of writing reveals how intertwined his religious commitments were with his reasoning.

Bacon, Francis. *The New Organon*. New York: Bobbs-Merrill, 1960. Part 1 is the account of what we think of as the modern experimental method. Important and very readable.

Baeyer, Hans Christian von. *Information: The New Language of Science*. New York: Phoenix Press, 2004. A short, clear, current account of the emergence of information as a feature of reality on a par with matter or even superior to it as the "stuff" of reality.

Baird, Davis. *Thing Knowledge: A Philosophy of Scientific Instruments*. Berkeley: University of California Press, 2004. A provocative monograph on the connections among ideas, the artifacts we call instruments, and knowledge claims.

Barabasi, Alberto-Laszlo. *Linked: The New Science of Networks*. Cambridge, MA: Perseus, 2002. This book is a joy to read, and without having to confront an equation, you will learn a great deal about how mathematical forms can have properties that affect physical systems—here, networks, which are, of course, relational structures. Very highly recommended.

Barber, Elizabeth Wayland. *The Mummies of Urumchi*. New York: Norton, 1999. Barber is a historian of ancient textiles, and this analysis of what we can learn from the clothing on a mummified, 3200-year-old Caucasian family found in central Asia is fascinating. Highly recommended.

Barbour, J. Murray. *Tuning and Temperament*. New York: Dover, 2004. A clear and correct account of the responses of musicians, composers, musical instrument makers, and philosophers to Pythagoras's claim that musical harmonies are mathematical. Recommended.

Bernoulli, Jacob. *The Art of Conjecturing*. Edith Dudley Sylla, trans. Baltimore: Johns Hopkins University Press, 2006. Sylla's introductory chapter to Bernoulli's 1709 text is a wonderful essay on the early history of applying mathematics to real-world decision-making. Recommended.

Bernstein, Peter. *Against the Gods: The Remarkable Story of Risk*. New York: Wiley, 1998. A well-written and highly informative popular history of practical applications of probability theory. Recommended.

Bertalanffy, Ludwig von. *General Systems Theory*. New York: George Braziller, 1976. A reprint of Bertalanffy's pioneering text on the systems idea. Dated, of course, but reading it puts you in touch with original thinking.

———. *Problems of Life*. New York: Dover, 1952. Here, Bertalanffy focuses on applying the systems concept in biology but also makes connections to other sciences.

Billington, David P. *The Innovators: The Engineering Pioneers Who Made America Modern*. New York: Wiley, 1996. An easy-to-read, insightful series of vignettes about the commercialization of key technological innovations in the 19th century. Recommended.

———, and David P. Billington, Jr. *Power, Speed and Form: Engineers and the Making of the Twentieth Century*. Princeton: Princeton University Press, 2006. Billington and his historian son extend the earlier model to eight key 20th-century innovations.

Biringuccio, Vannoccio. *Pirotechnia*. New York: Dover, 1990. Like Agricola's book, this is a 16th-century survey of technologies available to "engineers" of the period, focusing on uses of fire, heat, and explosives but ranging more widely.

Bolter, J. David. *Turing's Man*. Chapel Hill: University of North Carolina Press, 1984. Bolter was a humanist scholar responding to the very early stages of the computer's penetration of society, but his book remains a valuable reflection on a socially transformative technology.

Braudel, Fernand. *The Wheels of Commerce: Civilization and Capitalism, 15th–18th Century*. New York: Harper and Row, 1979. This is the second volume of a magisterial three-volume work of social history. Highly recommended.

Buchwald, Jed Z., and Andrew Warwick, eds. *Histories of the Electron*. Cambridge, MA: MIT Press, 2001. A first-rate professional history of science, this collection of essays examines in detail the discovery of the electron in 1897, the response of physicists to that discovery, and the impact of the electron on physics theory.

Carroll, Sean B. *Endless Forms Most Beautiful: The New Science of Evo Devo*. New York: Norton, 2005. Evolutionary development, the *Evo Devo* of the title, is a leading expression of Darwinian theory today, and this book describes it very well.

Cercignani, Carlos. *Ludwig Boltzmann: The Man Who Trusted Atoms*. Oxford: Oxford University Press, 1998. A new, very good biography of Boltzmann by a physicist; describes his contributions to physics, among them, to the physical reality of atoms. Recommended.

Ceruzzi, Paul. *A History of Modern Computing*. Cambridge, MA: MIT Press, 2003. A good history by a professional historian of technology.

Chandler, Alfred D., and James W. Cortada. *A Nation Transformed by Information*. Oxford: Oxford University Press, 2000. The authors examine the impact of information technologies on U.S. society.

Cline, Barbara Lovett. *Men Who Made a New Physics*. Chicago: University of Chicago Press, 1987. Very readable, short intellectual biographies of a handful of physicists who fashioned relativity and quantum theory in the first half of the 20th century. Recommended.

Cohen, I. Bernard. *The Birth of a New Physics*. New York: Norton, 1985. Cohen was a great historian of early modern science; this short monograph is an account of mechanics emerging as the foundation of physics in the 17th century. Recommended.

Cohen, Jack, and Ian Stewart. *The Collapse of Chaos*. New York: Viking, 1994. This lively book focuses on the maturation of chaos theory into complexity theory, carrying James Gleick's earlier book *Chaos* into the early 1990s.

Coleman, William. *Biology in the Nineteenth Century*. Cambridge: Cambridge University Press, 1977. An excellent, short history of major ideas at the heart of 19th-century biological theories. Highly recommended.

Cooper, Gail. *Air-Conditioning America*. Baltimore: Johns Hopkins University Press, 1998. A good history of the co-evolution of air-conditioning technology and its commercial, industrial, and residential applications.

Copernicus, Nicolaus. *On the Revolutions of the Heavenly Spheres*. New York: Prometheus, 1995. You can—and should—read book 1 of this epochal work to experience Copernicus's revolutionary idea firsthand. (The rest of the book requires serious effort!)

Cortada, James W. *Before the Computer*. Princeton: Princeton University Press, 2000. An interesting history of calculator and tabulator technologies in the 19th and early 20th centuries and how they affected the conduct of business and business strategies.

Cutcliffe, Stephen H., and Terry Reynolds. *Technology and American History*. Chicago: University of Chicago Press, 1997. An excellent collection of essays from the journal *Technology and Culture* on how technologies have changed American society. Recommended.

Darwin, Charles. *The Descent of Man*. London: Penguin, 2004. The extension of evolution to mankind came 12 years after *Origin* (and Wallace rejected it!). Highly recommended.

———. *On the Origin of Species*. Cambridge, MA: Harvard University Press, 1966. Reading Darwin's argument for what we call evolution is a powerful experience. Highly recommended, as is comparing it to Alfred Russel Wallace's essay "On the Tendency of Varieties to Diverge Indefinitely from Their Type" (download it from the Internet).

Debus, Alan. *Man and Nature in the Renaissance*. Cambridge: Cambridge University Press, 1978. A classic monograph in a wonderful series of history-of-science monographs by Cambridge University Press. Recommended.

Dennett, Daniel C. *Darwin's Dangerous Idea*. New York: Simon and Schuster, 1996. Provocative, even controversial, but Dennett is a leading philosopher and a defender of a strict, strictly atheistical reading of Darwinism.

Descartes, René. *Discourse on Method* and *Rules for the Direction of the Mind*. London: Penguin, 1999. These two short works are Descartes' prescriptions for a deduction-based methodology as the foundation for modern science.

———. *The Geometry*. New York: Dover, 1954. Title notwithstanding, this long essay, published in 1637, proposes substituting algebra for geometry as the basis of mathematics.

Dohrn-van Rossum, Gerhard. *History of the Hour*. Chicago: University of Chicago Press, 1996. An excellent history of the weight-driven clock and its impact on late-medieval and Renaissance society. Highly recommended.

Drachmann, A. G. *The Mechanical Technology of Greek and Roman Antiquity*. Madison: University of Wisconsin Press, 1963. A scholarly monograph reviewing just what the title promises. Out of print but available used.

Drake, Stillman, and I. E. Drabkin. *Mechanics in Sixteenth-Century Italy*. Madison: University of Wisconsin Press, 1969. The authors, leading historians of early modern science, give a detailed account of the state of the art in which Galileo was trained. Modest mathematics required, but very informative.

Edgerton, Samuel Y., Jr. *The Heritage of Giotto's Geometry: Art and Science on the Eve of the Scientific Revolution*. Ithaca: Cornell University Press, 1994. Out of print, but a good introduction to the idea that Renaissance painting techniques contributed to the rise of modern science. Highly recommended. Edgerton's *The Renaissance Discovery of Linear Perspective* covers the same material.

Einstein, Albert. *Relativity: The Special and General Theory*. London: Penguin, 2006. In his own words, writing for a general audience, Einstein describes relativity theory, aiming at a broad conceptual understanding. Highly recommended.

Eiseley, Loren. *Darwin's Century: Evolution and the Men Who Discovered It*. New York: Anchor, 1961. Eiseley wrote beautifully about science, and the virtues of this book include its clarity and readability. Recommended in spite of many more recent works on this topic.

———. *The Firmament of Time*. New York: Atheneum, 1960. Here, the writing is center stage. The theme is the naturalization of time and man in the 19th century. Highly recommended.

Eisenstein, Elizabeth L. *The Printing Revolution in Early Modern Europe*. Cambridge: Cambridge University Press, 2005. An important scholarly study of the social impact of print technology—this is an abridged and illustrated edition of Eisenstein's two-volume *The Printing Press as an Agent of Social Change* (1984)—attributing that impact primarily to features of the technology. Adrian Johns's book (below) takes issue with this view.

Eldredge, Niles. *Darwin: Discovering the Tree of Life*. New York: Norton, 2005. Eldredge is an important evolutionary biologist. Here, he traces the development of the ultimate unity of all life forms in Darwin's thinking.

Elkana, Yehuda. *The Discovery of the Conservation of Energy*. Cambridge, MA: Harvard University Press, 1974. A very good, nontechnical history of the idea of energy and the foundation of thermodynamics. Recommended.

Euclid. *Euclid's Elements*. Dana Densmore, ed. T. L. Heath, trans. Santa Fe, NM: Green Lion Press, 2002. You probably hated it in high school, but this is one of the truly great works of the mind, exemplifying reason and knowledge for mainstream Western intellectuals right down to the present. Read it to appreciate the mode of reasoning it exemplifies. Highly recommended.

Galen of Pergamon. *On the Natural Faculties*. New York: Putnam, 1916; in the Loeb Classical Library series and reprinted by Kessinger in 2004. In this work, Galen pulls together many strands of Greek and Graeco-Roman medical thought and theory, as Euclid did for Greek mathematics and Aristotle for Greek logic.

Galilei, Galileo. *Dialogue Concerning the Two Chief World Systems*, 2nd rev. ed. Berkeley: University of California Press, 1962. This is the book that caused Galileo's trial for heresy. Did he advocate the view that the Earth moved around the Sun? Is this a "fair" scientific treatment of a controversial issue? Highly recommended.

Gamow, George. *Thirty Years That Shook Physics*. New York: Anchor, 1966. Gamow remains one of the least known of the great 20th-century physicists. This is a wonderful autobiographical memoir, often funny and somewhat irreverent, of the creation of quantum mechanics. Very highly recommended.

Ghiselin, Michael T. *The Triumph of the Darwinian Method*. Chicago: University of Chicago Press, 1984. A prize-winning monograph on the logic of Darwin's argument in the *Origin* and his methodology. Recommended.

Gille, Bertrand. *Engineers of the Renaissance*. Cambridge, MA: MIT Press, 1966. Newer complement to Parsons (below) by a good French historian of technology; out of print but available used.

Gimpel, Jean. *The Medieval Machine: The Industrial Revolution of the Middle Ages*. New York: Holt, Rhinehart and Winston, 1976. Gimpel was an "amateur" historian in the best sense and an enthusiast for medieval and Renaissance technology as both beautiful and socially important. Easy to read yet packed with information. Highly recommended.

Gingrich, Owen. *The Book Nobody Read*. New York: Walker and Company, 2004. Gingrich is a leading historian of astronomy. Here, he traces the fate of copies of Copernicus's masterwork in the decades after its publication in 1543 as a way of assessing its influence.

Grafton, Anthony. *Leon Battista Alberti: Master Builder of the Italian Renaissance*. New York: Hill and Wang, 2000. Grafton is an intellectual historian, and this book, like his earlier biography of Jerome Cardan, gives an appreciation of the social-cum-intellectual context of a man who was at the center of art, business, and engineering in the late 16th century. Recommended.

Grant, Edward. *Physical Science in the Middle Ages*. Cambridge: Cambridge University Press, 1977. A very good, short monograph that surveys the major ideas, people, and places. A good source of leads to reading in greater depth about medieval nature philosophy as a seedbed of modern science.

Grattan-Guinness, Ivor. *The Norton History of the Mathematical Sciences: The Rainbow of Mathematics*. New York: Norton, 1997. Grattan-Guiness is the encyclopedic historian of mathematics, and this is a rich general reference to the subject.

Greene, Brian. *The Elegant Universe*. New York: Norton, 1999. All you want to know about string theory and more at the turn of the 21st century. Well and clearly written.

———. *The Fabric of the Cosmos*. New York: Norton, 2004. A popular treatment of late-20th-century cosmology by a Columbia University physicist. Very well written. Recommended.

Grendler, Paul F. *The Universities of the Italian Renaissance*. Baltimore: Johns Hopkins University Press, 2002. A scholarly study, narrow in scope, but this is the stuff of good history writing.

Hacking, Ian. *The Taming of Chance*. Cambridge: Cambridge University Press, 1991. Very good social-intellectual history of probability theory. Well written and, like all of Hacking's books, insightful. Recommended.

Hall, Marie Boas. *The Scientific Renaissance, 1450–1630*. New York: Dover, 1994. For today's historians of science, this is a dated book, but it is enjoyable to read, highly informative without being stuffy, and not wrong. A good lead-in to Westfall's *Construction* (below).

Hankins, Thomas L. *Science and the Enlightenment*. Cambridge: Cambridge University Press. An excellent, short book on science in the 18th century, with an emphasis on science as an agent of social reform through its connection to the ideas of rationality and progress. Highly recommended.

Harman, P. M. *Energy, Force and Matter: The Conceptual Development of Nineteenth-Century Physics*. Cambridge: Cambridge University Press, 1982. A very good, tightly focused book. Highly recommended.

Harris, William V. *Ancient Literacy*. Cambridge, MA: Harvard University Press, 1989. A valuable, detailed study of literacy in ancient Greece and Rome; the spread of writing in law, politics, and daily life; and the emergence of a commercial book trade. Recommended.

Harvey, William. *On the Motion of the Blood in Man and Animals*. New York: Prometheus, 1993. This remains one of the most readable of all primary source works in early modern science, arguing in 1628 for the circulation of the blood driven by the heart. Recommended.

Haskins, Charles Homer. *The Renaissance of the Twelfth Century*. Cambridge, MA: Harvard University Press, 2005. A classic essay, now reissued, that remains a joy to read as a general introduction to a field now dominated by specialists. Highly recommended.

———. *The Rise of Universities*. New York: Transactions, 2001. Again, a reissued early study. Recommended.

Hero of Alexandria. *Pneumatica*. New York: American Elsevier, 1971. Always in print, this little book, written in the 2nd century, is a collection of ideas for machines. A fascinating insight into one facet of Graeco-Roman technology, but note that the illustrations are not original. Recommended.

Hesse, Mary. *Forces and Fields: The Concept of Action at a Distance in the History of Physics*. New York: Dover, 2005. A valuable and insightful historical study of recourse by natural philosophers and modern scientists to forces acting at a distance, as opposed to mechanical contact forces, in order to explain natural phenomena without invoking magic or spiritualism. Highly recommended.

Hodges, Andrew. *Alan Turing: The Enigma*. New York: Simon and Schuster, 1983. An excellent biography of Alan Turing and a clear treatment of the intellectual context out of which the idea of the computer emerged. Recommended.

Holland, John H. *Hidden Order: How Adaptation Builds Complexity*. Boston: Addison-Wesley, 1996. Short monograph on self-organization by a pioneer of complexity theory and creator of the computer program Life. Highly recommended.

———. *Emergence: From Chaos to Order*. New York: Perseus Books, 1999. Holland gives scientific content to the cliché that the whole is greater than the sum of its parts. Recommended.

Hughes, Thomas P. *American Genesis: A Century of Invention and Technological Enthusiasm, 1870–1970*. Chicago: University of Chicago Press, 2004. Very good, very readable analysis of the relationships among technology, politics, and social values from the mid-19th through the mid-20th centuries. Recommended.

———. *Networks of Power*. Baltimore: Johns Hopkins University Press, 1983. Hughes is a dean of American historians of technology and this book shows why. It traces the relationships among invention, innovation, commerce, politics, science, and society in the creation of America's electrical networks. Recommended.

Johns, Adrian. *The Nature of the Book*. Chicago: University of Chicago Press, 1998. Johns argues at length in this big book, with lots of supporting detail, that print technology enabled authoritative, uniform versions of a text only after an extended social struggle to create institutions that protected profit and reduced pirated editions and plagiarism. Recommended.

Jones, Richard A. L. *Soft Machines: Nanotechnology and Life*. Oxford: Oxford University Press, 2004. The commercial exploitation of nanotechnology and that of molecular biology in the early 21st century are converging; this popular account of the convergence is timely.

Kelley, Donald. *Renaissance Humanism*. New York: Twayne, 1991. A good introduction to the Humanist movement by a major scholar. Recommended.

Klein, Jacob. *Greek Mathematical Thought and the Origin of Algebra*. New York: Dover, 1992. An original analysis and survey of Greek mathematics, which was, after all, a decisive influence on modern science.

Kramer, Samuel Noah. *Sumerian Mythology*. Philadelphia: University of Pennsylvania Press, 1972. Kramer was one of the pioneers of the study of Sumerian texts. This book presents deciphered Sumerian religious texts.

Kuhn, Thomas. *The Copernican Revolution: Planetary Astronomy in the History of Western Thought*. Cambridge, MA: Harvard University Press, 1992. An excellent, important analysis of the intellectual legacy of Copernicus's astronomical ideas. Highly recommended.

Lamarck, Jean-Baptiste. *Zoological Philosophy*. Chicago: University of Chicago Press, 1976. Lamarck was a far more important figure than most 20th-century biologists are willing to allow. Read this for yourself and see that Lamarck is more than just the inheritance of acquired characteristics!

Landels, J. G. *Engineering in the Ancient World*. London: Chatto and Windus, 1978. Easier to find than Drachmann's book (above), though also out of print. Both are the real thing: Engineering-knowledgeable authors use surviving texts and artifacts to reveal what the Graeco-Romans knew how to do with machinery. Sounds stuffy, but it's fascinating detective work.

Lefevre, Wolfgang. *Picturing Machines, 1400–1700*. Cambridge, MA: MIT Press 2004. A collection of essays that survey the evolution of machine drawing during the Renaissance and its implications for machine design and construction and, more broadly, for technological innovation as a social force. Recommended.

Levere, Trevor. *Transforming Matter*. Baltimore: Johns Hopkins University Press, 2001. Histories of chemistry are rare, and readable ones (to non-chemists), rarer still; thus, Levere's book is recommended.

Lewontin, Richard. *The Triple Helix*. Cambridge, MA: Harvard University Press, 2000. Lewontin is a major figure in evolutionary biology, and in these four lectures, he describes what he thinks is wrong with the current linkage of evolution to molecular biology and genetics. Recommended (but be prepared for its relentless negativism!).

Lindberg, David C. *The Beginnings of Western Science*. Chicago: University of Chicago Press, 1992. A fine example of history-of-science writing and scholarship, surveying the Greek, Roman, Islamic, medieval, and early Renaissance antecedents of modern science. Recommended.

Lindley, David. *Boltzmann's Atom*. New York: Free Press, 2001. Where Cercignani's book focuses on Boltzmann and his scientific accomplishments, Lindley focuses on the idea of the atom in the 19th century and the context within which Boltzmann championed its reality.

Lloyd, Seth. *Programming the Universe*. New York: Knopf, 2006. Lloyd is a pioneer of the quantum computer and here describes, in nontechnical terms, his conception of the Universe as a quantum computational information structure. Recommended.

Lucretius. *On the Nature of the Universe*. London: Penguin, 1994. An important and, from the Renaissance on, influential statement of Epicurus's atomism that made the armature of a philosophy of nature and of man.

Mandelbrot, Benoit. *The Fractal Geometry of Nature*. San Francisco: W.H. Freeman, 1983. Mandelbrot pioneered the field of fractional dimensionality. This is a stimulating, accessible description of what fractals are and why they matter. Subsequently, they have become important tools in applied mathematics. Recommended.

Mann, Charles C. *1491: New Revelations of the Americas Before Columbus*. New York: Knopf, 2005. Mann is a journalist, synthesizing primary source material, and his claims are controversial, but they reflect the opinions of a growing number of scholars that the inhabitants of the Americas before 1492 were far more numerous and far more sophisticated than we have been taught.

Marenbon, John. *Later Medieval Philosophy*. London: Routledge, 1991. A good introduction to the knowledge issues in medieval philosophy, but this source also describes the rise of the universities and the translation of Greek and Roman texts from Arabic into Latin. Recommended.

Mayr, Ernst. *The Growth of Biological Thought*. Cambridge, MA: Harvard University Press, 1982. Mayr was one of the great evolutionary biologists of the 20th century and was still publishing as he approached 100! This is an excellent history of the great 19th-century ideas in biology. Recommended.

McClellan, James E., III, and Harold Dorn. *Science and Technology in World History: An Introduction*. Baltimore: Johns Hopkins University Press, 2006. Excellent social-historical interpretation of how technology and, later, science-through-technology have changed the world. Recommended.

Melsen, A. G. van. *From Atomos to Atom: The History of the Concept Atom* New York: Harper, 1952. A dated but charmingly literate monograph on the atomic idea from the Greeks to the 20th century. Out of print but available used. See Pullman below.

Misa, Thomas J. *A Nation Transformed by Steel*. Baltimore: Johns Hopkins University Press, 1995. An excellent example of the placement of history of technology in its social context. The displacement of iron by steel implicated science, technology, finance, industry, government, and society, and Misa does justice to them all. Highly recommended.

Morange, Michel. *A History of Molecular Biology*. Cambridge, MA: Harvard University Press, 1998. If you're going to read just one book about molecular biology, make it this one. Morange is a biologist, not a historian, so the focus is on the science, but the writing makes it very accessible.

———. *The Misunderstood Gene*. Cambridge, MA: Harvard University Press, 2001. Highly recommended critique of the still-prevalent view that genes do it all. What genes are and how they act is still being discovered.

Newton, Isaac. *The Principia*. I. Bernard Cohen and Anne Whitman, trans. Berkeley: University of California Press, 1999. This is one of the most influential science texts ever published, and in this new translation with extensive commentary, the general reader can learn directly from Newton! Highly recommended.

Nisbet, Robert. *A History of the Idea of Progress*. Piscataway, NJ: Transaction, 1994. Intellectuals have been critical of the idea of progress for most of the 20th century, but it remains a core public value and central to science and technology. Nisbet's book takes a positive view of progress and is a successor to J. B. Bury's classic *The Idea of Progress*. Recommended.

Nye, Mary Jo. *Before Big Science: The Pursuit of Modern Chemistry and Physics, 1800–1940*. New York: Twayne, 1996. A good short history of physical science as it became a driver of social change. Highly recommended.

Overbye, Dennis. *Lonely Hearts of the Cosmos*. Boston: Little, Brown and Co., 1999. A wonderful book that uses people, their ideas, and relationships as a means of describing the development of ideas about the origin of the Universe since the 1950s. Highly recommended. (Read the revised 1999 edition or a later one.)

Parsons, William Barclay. *Engineers and Engineering in the Renaissance*. Cambridge, MA: MIT Press, 1968. A classic study of what the title promises. Still good enough after its original publication in the late 1930s to remain in print for decades, and now available used at reasonable prices. A big book.

Pesic, Peter. *Abel's Proof.* Cambridge, MA: MIT Press, 2003. Very short, very good (and nontechnical) account of how Niels Abel's proof of a negative about algebraic equations led to major innovations in 19th-century mathematics and in late-20th-century symmetry-based physics.

Plato. *Plato: The Collected Dialogues.* Edith Hamilton and Huntington Cairns, eds. Princeton: Bollingen, 1978. The dialogue *Phaedrus* contains Socrates's argument against writing; the dialogues *Thaeatetus* and *Timaeus* relate to knowledge of nature.

Porter, Roy. *The Rise of Statistical Thinking, 1820–1900.* Princeton: Princeton University Press, 1986. An excellent book; nicely complements Hacking (above), with a narrower focus.

Prager, Frank D., and Gustina Scaglia. *Brunelleschi: Studies of His Technology and Inventions.* New York: Dover, 2004. Brunelleschi not only reintroduced perspective drawing, but his dome for the cathedral in Florence was an epochal technological achievement, and he invented numerous machines to enable its construction. Recommended.

Prigogine, Ilya. *Order Out of Chaos.* New York: Bantam, 1984. A philosophical reflection on the challenge of process thinking to atomistic thinking. Highly recommended. (His later *From Being to Becoming* is more challenging technically.)

Provine, William. *The Origins of Theoretical Population Genetics.* Oxford: Oxford University Press, 2003. Ignore the forbidding title: This is an excellent book that exposes how Darwinian evolutionary theory was resurrected in the 1920s. Highly recommended.

Pugsley, Alfred, ed. *The Works of Isambard Kingdom Brunel.* Cambridge: Cambridge University Press, 1976. Brunel, typically for engineers, is unknown in spite of being one of a small community of men (including his father, Marc) responsible for making the world "modern." This is a collection of short essays (search for it used online) that reveal how much one of these men accomplished while knowing so little theory!

Pullman, Bernard. *The Atom in the History of Human Thought.* Alex Reisinger, trans. Oxford: Oxford University Press, 2001. A history of the atom from the Greeks to the 20th century. Given that the author was a professor of chemistry at the Sorbonne, this is a scientist's view of the history of a core scientific idea.

Raby, Peter. *Alfred Russel Wallace: A Life.* Princeton: Princeton University Press, 2001. A good biography of Wallace, who is still damned with faint praise by biologists. A leading scientist, a social reformer, and a spiritualist, Wallace deserves our attention. Recommended.

Randall, Lisa. *Warped Passages*. New York: Harper, 2005. If you're interested in learning about string theory, this is one breezily written option by a string theory researcher. I prefer Brian Greene's *The Elegant Universe* on this subject.

Ratner, Mark, and Daniel Ratner. *Nanotechnology: A Gentle Introduction to the Next Big Idea*. Upper Saddle River, NJ: Prentice Hall, 2002. Nanotechnology research, development, and commercialization, along with safety and health issues, are evolving at a breakneck pace, so consider this a good introduction to the underlying ideas and read the newspaper.

Robb, Christina. *This Changes Everything: The Relational Revolution in Psychology*. New York: Farrar, Straus and Giroux, 2006. Robb is a Pulitzer Prize–sharing journalist; this book describes how acknowledging the reality and causal efficacy of relationships affected clinical and theoretical psychology.

Rocke, Alan J. *Chemical Atomism in the 19th Century: From Dalton to Cannizaro*. Columbus: Ohio State University Press, 1984. An in-depth study of the early history of the atomic theory of matter, when it was mostly a theory for chemists.

Rudwick, Martin J. S. *The Meaning of Fossils*. Chicago: University of Chicago Press, 1985. All of Rudwick's books are excellent, and his most recent, *Bursting the Limits of Time*, is most relevant to the reconceptualization of time in the 19th century, but it is massive. This book is a gentler yet still highly informative study of the same subject. Highly recommended.

Scaglia, Gustina. *Mariano Taccola and His Book* De Ingeneis. Cambridge, MA: MIT Press, 1972. This is a very good edition, with scholarly commentary, of a Renaissance-era machine design book, symptomatic of the emergence of modern engineering. See Prager and Scaglia (above) and Scaglia's *Francesco di Giorgio*, a beautiful collection of Renaissance machine drawings with extensive commentary by Scaglia. Recommended.

Seife, Charles. *Decoding the Universe*. New York: Viking, 2006. A very readable account by a science journalist of how information has become physically real for many scientists. Recommended.

Shapin, Steven and Simon Schaffer. *Leviathan and the Air Pump: Hobbes, Boyle and the Experimental Life*. Princeton, New Jersey: Princeton

University Press, 1985. A modern classic that exposes the equivocal character of experimental research using newly devised instruments by way of Thomas Hobbes' criticism of Robert Boyle's "discoveries" and the Royal Society as an institution. Recommended.

Singleton, Charles. *Art, Science, and History in the Renaissance*. Baltimore: Johns Hopkins University Press, 1968. An alternative to Edgerton (above); also out of print but available used at more reasonable prices.

Smolin, Lee. *The Trouble with Physics: The Rise of String Theory, the Fall of Science, and What Comes Next*. Boston: Houghton Mifflin, 2006. One of several recent attacks on string theory by physicists who claim that it is a dead end and bad science. I admire Smolin's earlier books, especially *Three Roads to Quantum Gravity*, and his criticisms are legitimate, but the book's primary value is as one skirmish in a "war" within physics.

Sorabji, Richard. *Matter, Space and Motion: Theories in Antiquity and Their Sequel*. Ithaca: Cornell University Press, 1988. A survey by a philosopher of materialist theories of nature from the earliest Greek philosophers through the 6th century. This complements Lindberg's book, above.

Stachel, John. *Einstein's Miraculous Year*. Princeton: Princeton University Press, 1998. Stachel is the editor of the Einstein papers and here offers the historic 1905 articles in translation with enough commentary for anyone to follow their arguments. Highly recommended.

Stephenson, Bruce. *The Music of the Heavens: Kepler's Harmonic Astronomy*. Princeton: Princeton University Press, 1994. Here, you can see the authority given to the Pythagorean idea that mathematical form was the indwelling order of nature underlying its expression in matter and that this order was fundamentally musical. Recommended.

Strogatz, Steven. *SYNC: The Emerging Science of Spontaneous Order*. New York: Hyperion, 2003. A readable book for a general audience on self-organization, bringing Prigogine's early ideas up-to-date. Strogatz is himself a researcher in this field.

Susskind, Leonard. *The Cosmic Landscape: String Theory and the Illusion of Intelligent Design*. Boston: Little, Brown and Co., 2006. As if to counter Smolin's rant against string theory, one of its architects describes the theory as if its triumphant completion and confirmation were imminent! Stranger than science fiction. Recommended.

Swade, Dorn. *The Difference Engine*. London: Penguin, 2002. An excellent book that makes reading about Charles Babbage's failed quest to build a computer in the mid-19th century fun. The effort and its failure reveal much about the science-technology-society relationship. Highly recommended.

Taylor, George Rogers. *The Transportation Revolution*. New York: Harper, 1968. A solid history of the 19th-century transportation technology innovations that literally changed the world.

Travis, Anthony. *The Rainbow Makers*. Bethlehem, PA: Lehigh University Press, 1983. The fascinating story of how a teenager discovered the first synthetic dye and triggered the first "Silicon Valley" phenomenon, in which chemical science, industry, and government created enormous wealth and power. Recommended.

Uglow, Jenny. *The Lunar Men: Five Friends Whose Curiosity Changed the World*. New York: Farrar, Straus and Giroux, 2002. Wonderful account of the interactions of a group of thinkers and doers at the turn of the 19th century whose members included James Watt and Mathew Boulton, Erasmus Darwin and Josiah Wedgwood (both of them Charles Darwin's grandfathers!), and Joseph Priestley. Highly recommended.

Vitruvius. *The Ten Books on Architecture*. New York: Dover, 1960. More than 2000 years old, still in print, and still worth reading!

Watson, James D. *DNA*. New York: Knopf, 2003. Now an honored senior, Watson describes the state of our understanding of how DNA works for a general audience. Recommended.

———. *The Double Helix*. New York: Signet, 1968. No scientist had written such a tell-it-all account of how his research was done before this book. Highly recommended.

Webster, Charles. *The Great Instauration*. London: Peter Lang, 2002. Webster describes the social context of 17th-century England—especially the religious, political, social reform, and medical contexts—in which early modern science took root.

Weinberg, Steven. *Dreams of a Final Theory*. New York: Pantheon, 1992. Weinberg shared a Nobel for the first step in the unification of the four fundamental forces of nature and here anticipates the implications of full unification. Not dated because there has been little progress since 1992!

Westfall, Richard S. *The Construction of Modern Science*. Cambridge: Cambridge University Press, 1989. Short, excellent introduction to the ideas at the heart of 17th-century science and its accomplishments. Highly recommended, as are all the monographs in the Cambridge History of Science series.

———. *Never at Rest: A Biography of Isaac Newton*. Cambridge: Cambridge University Press, 1980. The definitive personal and intellectual biography of Newton. Highly recommended.

White, Lynn. *Medieval Technology and Social Change*. Oxford: Oxford University Press, 1966. White describes the social impact of the stirrup, wind power, and agricultural innovations, overstating the case but calling attention to technology as a force driving social change when most historians ignored it.

Williams, Trevor. *A History of Invention*. New York: Checkmark Books, 1987. It looks like a coffee table book, but Williams is a scholar and the book is filled with lots of valuable information without reading like an encyclopedia.

Wilson, Catherine. *The Invisible World: Early Modern Philosophy and the Invention of the Microscope*. Princeton: Princeton University Press, 1995. An important study of the interaction of ideas and instruments, theories of nature and observations. Recommended.

Worboys, Michael. *Spreading Germs: Disease Theories and Medical Practice in Britain, 1865–1900*. Cambridge: Cambridge University Press, 2000. An account of the response of the British medical community to the germ theory of disease as that theory evolved. Recommended.

Zachary, G. Pascal. *Endless Frontier: Vannevar Bush, Engineer of the American Century*. Cambridge, MA: MIT Press, 1999. A good biography of the man who was the "czar" of harnessing science and technology to the World War II effort and who promoted the postwar policy of federal support for scientific research.

Zagorin, Perez. *Francis Bacon*. Princeton: Princeton University Press, 1998. A very good biography of Bacon, doing justice to him as a social reformer, political opportunist, and philosopher of nature. Recommended.

Internet Resources:

Ancient Languages and Scripts. An informative site on the history of writing. www.plu.edu/~ryandp/texts.html.

The Art of Renaissance Science. A rich, multi-disciplinary site created by Joseph Dauben on the relationships among art, mathematics and science in the Renaissance. www.mcm.edu/academic/galileo/ars/arshtml/arstoc.html.

The Galileo Project. An excellent resource site for everything to do with Galileo's life, works and ideas. http://galileo.rice.edu.

The History of Computing. An excellent collection of materials and links for exploring the history of computers and computing. http://ei.cs.vt.edu/~history/.

The Labyrinth: Recourses for Medieval Studies. A "super" site listing resources for exploring Medieval culture. http://labyrinth.georgetown.edu.

NASA History Division. A central site for aerospace history. http://history.nasa.gov.

National Nanotechnology Initiative. The official Web site for federally funded nanotechnology research and development. www.nano.gov.

The Newton Project. A similar, and similarly excellent resource, for the life, works and ideas of Isaac Newton. www.newtonproject.ic.ac.uk.

The Nobel Foundation. Official web site offering access to all Nobel Prize winners, their biographies and accomplishments, and their acceptance addresses; a rich and fascinating history of science resource. http://nobelprize.org.

The Official String Theory Website. The "home page" for accessible accounts of string theory. http://superstringtheory.com.

Selected Classic Papers from the History of Chemistry. An outstanding collection of the full text of classic papers in the history of chemistry. http://web.lemoyne.edu/~giunta/papers.html.

Stanford Encyclopedia of Philosophy. A superb resource for the history of philosophy, of uniformly high quality, guaranteed to illuminate and please. Includes outstanding entries on many science topics—try advanced search. http://plato.stanford.edu.

Sunny Y. Auyang. "Scientific convergence in the birth of molecular biology." Very good essay on the history of molecular biology. Other articles available on this idiosyncratic yet interesting website by a respected scientist address engineering, including a useful history of engineering, biomedicine, and physics. www.creatingtechnology.org/biomed/dna.htm.

University of Delaware Library. *Internet Resources for History of Science and Technology*. A "super" site for exploring the history of technology and of science from antiquity to the present. www2.lib.udel.edu/subj/hsci/internet.html.

University of St. Andrews School of Mathematics and Statistics, *The MacTutor History of Mathematics Archive*. A very good resource for the history of mathematics; the Biographies on the site offer a comprehensive history of mathematicians and their accomplishments. www-history.mcs.st-and.ac.uk/.

Notes

Notes

Notes

Notes